Failure Analysis in
Engineering Applications

Failure Analysis in Engineering Applications

Shin-ichi Nishida Dr Eng

Butterworth-Heinemann Ltd
Linacre House, Jordan Hill, Oxford OX2 8DP

 PART OF REED INTERNATIONAL BOOKS

OXFORD   LONDON   BOSTON
MUNICH   NEW DELHI   SINGAPORE   SYDNEY
TOKYO   TORONTO   WELLINGTON

First published by the Nikkan Kogyo Shimbun, Ltd. 1986
First published in Great Britain by Butterworth-Heinemann Ltd 1992

British Library Cataloguing in Publication Data
Nishida, Shin-ichi
    Failure analysis in engineering applications.
    I. Title
    620.1

ISBN 0 7506 1065 4

Library of Congress Cataloguing in Publication Data
Nishida, S. (Shin-ichi)
    Failure analysis in engineering applications/ Shin-ichi Nishida.
        p.   cm.
    Translation from Japanese.
    ISBN 0 7506 1065 4
    1. System failures (Engineering)   2. Fracture mechanics.
    I. Title.
    TA169.5.N57   1991
    620'.0042–dc20   91–17844   CIP
Printed and bound in Great Britain by Thomson Litho Ltd., East Kilbride, Scotland
Typeset by Vision Typesetting, Manchester

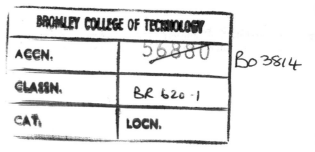

Contents

Foreword

With the recent trend for sophistication and enlargement of machinery and structures, strength requirements have become more and more rigorous. Against this background, the author describes various strength-related items, in terms of safety and economy, for the rational design and application of machinery and structures. This book is based on the author's abundant field experience and reveals a profound insight. It gives examples of failure and suggests possible countermeasures.

After obtaining his doctorate at Kyushu University, the author joined Nippon Steel Corporation in April 1970. Since then, he has devoted himself to solving problems related to the strength of steel products and to studying fracture mechanics. Many researchers have given their attention to his development of the unique High-Speed Rail Testing Machine, awarded the prize of the Engineering Division of the Japan Society of Mechanical Engineers in 1985, and designed to use the existing rails of the Shinkansen (the main line of the former JNR), as well as to his solution of practical problems through experiments using the testing machine.

The book is written from a broad point of view, for designer and engineers who actually carry out the design and maintenance of machinery and structures.

I am confident that the book will also prove valuable for students who are beginning a study of the subject of fracture problems.

Hironobu Nisitani, Dr Eng.
Professor, Department of Mechanical Engineering,
Kyushu University

Preface

There is a proverb which says, 'Learn wisdom by the follies of others'. This can be interpreted as a lesson that before we laugh at someone else's behaviour, we should change our own. If this is applied to things or to equipment instead of to a person, how do we then interpret the proverb? We often have the experience of observing the failure or poor condition of some facility. Unless a facility is very special, it is not unusual to have several similar facilities installed in the same company or organization. When one of these facilities fails, it may be possible to prevent similar failures by investigating and analysing the initial failure in detail. If this is possible, the advantage provided will be immeasurable in terms of the effective use of limited resources. This book was written from this viewpoint. Because there is still much to be clarified in the author's study and because the range of the analysis of the applied questions is very wide, the reader may be dissatisfied with the book in many respects.

However, the author has decided to publish the book, showing some actual examples, in spite of these imperfections, on the assumption that some insufficiencies will be tolerated or modified by the reader. This kind of book could never be published if full marks are aimed at (see Figure A).

Conventional engineering units are used in this book, because they are still used by most engineers and designers to relate to machinery and equipment, and by researchers who are interested in the strength of materials. To convert conventional engineering units into SI units see Table A1.1 in Appendix 1.

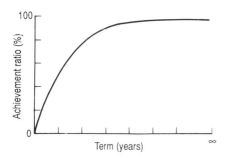

Figure A Relationship between term and achievement ratio

Chapter 1 describes fundamental items relating to failures, and states that a failure results not only in direct losses, but also in indirect losses, and that in some cases the image of the enterprise or organization in question is damaged by the failure. The importance of failure analysis is also described.

Chapter 2 deals with a procedure for failure analysis. In the procedure for failure analysis shown, a method for removing metal deposits (rust) is described, because the usual structural members are often exposed to corrosive environments during their long-term service. In addition, literature useful for failure analysis is listed.

Chapter 3 describes basic knowledge and data necessary for analysing various examples of failure.

Chapter 4 describes more than 40 examples of failure and shows how to analyse all the failures in full detail. The examples include failures often experienced in daily life, those which inevitably occur some day, as with wire ropes, failures of bolts and gears that are typical machine parts, and those which seem to be strange phenomena and can be understood only after clarification, as in the examples of piping valves and a pipeline that failed before the start of operation. In most components, failure is initiated in a stress concentration area such as a notch. Failures in transmission shafts and welds that are typical of this type of failure are also shown.

The author is indebted to the many people both within and outside the company for their guidance and encouragement in the completion of this book. In particular, the author extends his sincere thanks to Professor H. Nisitani, Department of Mechanical Engineering, Kyushu University, who has guided him for 20 years since his student days. The author is also grateful to Professor Emeritus F. Hirano, Department of Mechanical Engineering, Kyushu University; Mr Yukio Nagao, the former chief researcher, Mitsui Engineering and Shipbuilding Company, Ltd; and Mr Yoshikazu Sano, Engineering Manager, Hitachi Metals, Ltd, for giving permission to quote their work.

The author is also grateful to the following Nippon Steel members who provided willing cooperation: Mr Shigeo Hosoki, Executive Vice President and Managing Director of the Central R&D Bureau; Mr Hiroki Masumoto, Chief Researcher; Mr Mitsuo Kanda,

General Manager of Yawata R&D Laboratory; Dr Kazuo Sugino, Chief Researcher; Mr Chikayuki Urashima, Researcher; Mr Akira Yagi, Researcher; Mr Shigeru Mizoguchi, Researcher; and Mr Takayuki Yonekura, Manager. Mr Chikayuki Urashima has cooperated with the author as a partner and this book would not otherwise have been completed. The author thanks all those who have supplied him with good suggestions every day, especially the people of the Central R&D Bureau of head office, Technical Administration Bureau, Technical Planning and Development Coordination Division, Yawata Works, Plant Engineering and Technology Bureau, and Tobata Plant and Machinery Works.

In addition, the author thanks Professor A.J. McEvily, Department of Metallurgy, University of Connecticut, USA, for his timely recommendation to publish and Professor R.A. Smith, Department of Mechanical and Process Engineering, Sheffield University, UK, for his kind introduction to the publisher, as well as Mr Andrew Burrows, Commissioning Editor and his co-workers in Butterworth Scientific Ltd.

Shin-ichi Nishida
October 1986
(Translated into English, May 1990)

1 Strength of materials and kinds of failure

1.1 Importance of failure analysis

Side by side with recent advances in industrial technology and the pursuit of high efficiency and productivity have come requirements such as cost reduction, necessary economies in energy and labour, and process omission. Moreover, an increase in the size of steel structures, together with demands for improved performance, bring requirements for service environments that are becoming more and more rigorous. However, with the development of the social sciences, the greatest concern has been with the safety of equipment. Therefore, today's designers must design from the mutually contradictory standpoints of high performance, cost reduction and high reliability. Design bases can be broadly classified into the following categories:

1. Design that does not allow failures (safe-life design) and
2. Design that allows failures but ensures safety through maintenance (damage-tolerant design or fail-safe design).

Almost all steel structures are designed on the basis of category 1 above. Case 2 is limited to aircraft with severe weight limits, and components that have finite service life in terms of service characteristics, such as bearings and wire ropes, etc.

Many reports describe instances where structures have failed under quite low loads although they were designed to be failure free, with generously high estimated values of safety factor (S. Nishida, C. Urashima and H. Masumoto, 1976, 1979 and 1980, unpublished report) [1–5]. As is well known, more than 80–90% of failures in steel structures are said to be directly or indirectly caused by fatigue. These figures are in fact supported by the results of investigations of failures conducted by the author and his cooperating researchers (S. Nishida, C. Urashima and H. Masumoto, 1976, 1979 and 1980, unpublished report) [6]. Partly because of this there are many fatigue researchers in Japan and a number of people attach much importance to the failure phenomenon. In research in this field, however, emphasis so far has been laid on the clarification of fundamental phenomena, and little progress has been made in practical research, in particular because of the large number of factors relating to fatigue failure. Moreover, almost all examples of failure bring dishonour to those concerned, and those who investigate the failures feel that they are engaged in backward-looking research. For these reasons, failures have rarely been publicized and it is not too much to say that the people concerned are not ready positively to utilize examples of failures as useful lessons.

Examples of failure can be valuable lessons in how to prevent failure. In general, examples of failure represent data under actual conditions that cannot be obtained from small-scale experiments in the laboratory. It is beyond dispute that failure stories are generally far more useful than success stories. It is necessary to clarify the causes of various failures and to take countermeasures to prevent the recurrence of similar failures. This is also very important because the results can be used to determine effective design guidelines. In this book emphasis is thus laid on the importance of failure analysis, and some examples of the method of analysis are shown. Many of the examples of failure analysis are related to failures caused by fatigue. However, examples of failure ascribed to other causes are also provided. The author considers that universal application of the examples of failure and countermeasures shown in this book is possible through an understanding of the basic concept of failure analysis.

1.2 Failure can cause heavy losses

When a failure occurs, some losses, large or small, will be generated. In particular, when an entire plant or company works on a computer system, trouble in one location will affect all lines. Therefore, when a failure has occurred, it is necessary to clarify the cause of the failure and to take appropriate countermeasures. For this reasons, failure is defined first in this section and losses caused by failures are then described.

1.2.1 Definition of failure

Failure is a general term for a condition in which a member is subjected to plastic deformation; in other words, where irreversible traces are observed in a member. Failures can be broadly classified into the following four categories:

1. damage
2. fracture
3. break, and
4. rupture.

Damage means a condition in which some plastic flow accumulates in a member compared with the brand-new condition, and in this damaged condition the member can still be used. Fracture indicates a state in which a crack has been initiated. Break represents a state in which a member is separated into two or more parts. However, fracture may be used to indicate the condition of a break (for example, a fracture surface) or is often used in the same sense as for the cut-off or disconnection of wire products. Rupture indicates a special case of break and is used to indicate the breakage of a very ductile material by plastic slip (for example, creep rupture). It is impossible, or almost impossible, to use a member in a state of fracture or break as it is. In some cases, damage in a broad sense is used in the same sense as failure. The definition of failure is summarized in Figure 1.1.

In the static tensile test, failure occurs when a stress exceeding the yield stress is applied. In the case of fatigue, however, plastic deformation often occurs in a member locally and partially. The definition shown in Figure 1.1 therefore applies to the macroscopic phenomenon. In other words, a fatigue crack is initiated under a stress below the elastic limit and may propagate to develop into final fracture. Therefore, failure, damage and fracture can occur under stress even below the yield stress of a member.

Damage in a broad sense includes fatigue, wear and corrosion. As shown in Table 1.1 [6], damage due to fatigue is macroscopically invisible and proceeds

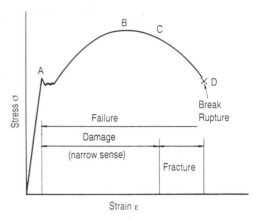

Figure 1.1   Definition of failure

rapidly; this kind of damage is therefore dangerous in many instances. However, damage due to wear and corrosion is macroscopically visible and proceeds gradually; the safety of a member can be easily checked by conducting intermediate inspection in this instance. Moreover, sudden break may sometimes occur because of impact loads or overloads. Most of these failures are caused by human error in operation, design or manufacture; these failures cannot be neglected. Failures mainly caused by fatigue and sudden failures under static loads are therefore mainly described from the viewpoint of frequency of occurrence and safety.

1.2.2 Losses through failures

Losses caused by failures are shown in Figure 1.2. As described previously, more than 85–90% of failures are generally caused directly or indirectly by fatigue. Losses through failures are classified as direct and indirect. The former includes the cost of repair work, the cost of work to prevent failures, accident compensation, etc. It is desirable to carry out repair work and work to prevent failures at the same time. In the case of production facilities, however, work to prevent failures is often conducted later during the scheduled shutdown period by giving top priority to restoration, because a line stoppage results in a decrease in

Table 1.1   *Types of failure (in broad sense) and macrofeatures* [6]

Type	Frequency of generation	Macroscopic phenomenon	Macroscopic growth	Safety
Fatigue	1	Invisible	Rapid	Dangerous
Wear	2	Visible	Gradual	Safe
Corrosion	3	Visible	Gradual	Safe
Others (impact load, overload)	4	Invisible	Rapid	Dangerous

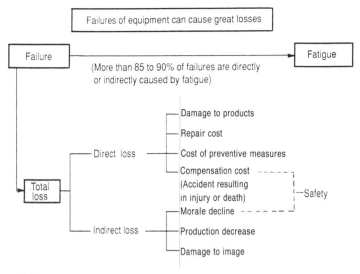

Figure 1.2 Failures and losses

production. When a failure results in the injury or death of workers or causes damage to people in the neighbourhood, the problem of accident compensation occurs. Furthermore, such failures may bring about an on-the-spot inspection by the authorities concerned and a responsibility problem for the plant safety supervisor. To prevent such an occurrence, efforts to prevent failures and to carry out maintenance work should be made every day. Indirect losses include a decrease in production, and damage to the company's image. We often hear that operators are frightened of increasing the output of a plant for some time immediately after a major accident even if the distance between the control room and the plant is sufficiently great. Although importance has been attached to direct losses in the past, due consideration has recently been given to damage to the image of a company, which prevents the recruitment of first-class workers.

In an extreme case, the fracture of a bolt in a modern computerized system may result in a line stoppage [7]. Therefore, it is necessary to investigate even minor failures as thoroughly as possible instead of making repairs according to some haphazard policy, and there are signs of attempts at a wider contribution to society by not only preventing the recurrence of similar failures, but also by making public the results of such investigations (S. Nishida, C. Urashima and H. Masumoto, 1976, 1979 and 1980, unpublished report) [2–6] (see also Section 2.4.3). When the failure investigation is conducted by both the designer and the user, the cause of a failure can be clarified more thoroughly and it is easy to adopt ideal measures.

1.3 Conditions for the occurrence of failures, and causes of failures

1.3.1 Conditions for the occurrence of failures

Figure 1.3 shows the conditions for the occurrence of failures and the countermeasures to prevent failures. Needless to say, a failure occurs when an external force exceeds the resistance of the material. It is necessary to examine whether the cause of the failure is a miscalculation of the external force or the wrong selection of a material. Truly rational measures can be taken by examining factors related to both causes.

In other words, it is necessary to take measures to reduce the external forces (including stress) or to increase the resistance or to take measures related to both. In general, those designers who are strong on mechanics may be weak on materials, and vice versa. Therefore, there are many cases in which design is very uneconomical. It is necessary to consider the two factors, i.e. external forces and materials, in determining countermeasures to prevent failures, and to draw a truly rational conclusion on the basis of the effect, cost, etc. of the measures.

1.3.2 Damaged members and a breakdown of causes of failures

The details of the failures of machines and mechanical parts which the author and his cooperating researchers investigated are shown in Figure 1.4 [6]. As is apparent from the figure, the greatest number of failures is observed in welds, and failures other than

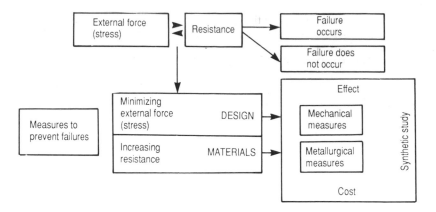

Figure 1.3   Occurrence of failures and measures to prevent them

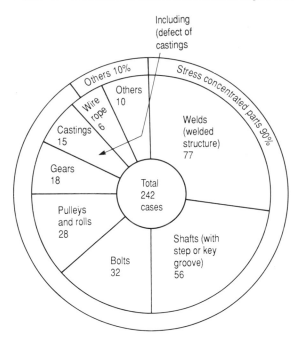

Figure 1.4   Classification of failures according to failed members

welds decrease in the following order: shafts, bolts, pulleys or rolls, gears, wire ropes, etc. The reason why the greatest number of failures is observed in welds may be because almost all assembly parts are welded parts, the absolute number of welded parts is large, and the strength of welds is generally lower than that of the base metal. The reason why the number of failures in shafts is the second largest after that of welds may be that a shaft is an important member for transmitting power and that replacements cannot be easily or quickly procured.

Although bolts rank third, it might be thought that bolts have, in reality, the largest absolute number of failures among mechanical parts. However, it seems that most broken bolts are appropriately replaced on site because, except in special instances, bolts can be procured soon after breakage or can be fabricated easily. Like shafts, pulleys, rolls, gears and wires are indispensable for transmitting loads, and they are typical mechanical parts. These failures are classified by causes, as shown in Figure 1.5. As is apparent from Figure 1.5, more than 80% of failures are caused by

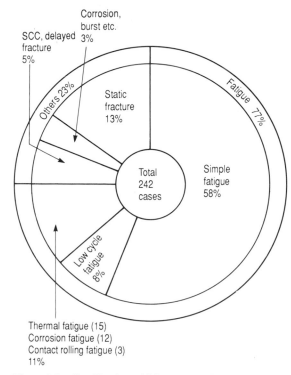

Figure 1.5   Classification of failures according to cause

fatigue (including simple fatigue, corrosion fatigue, thermal fatigue, etc.), and other causes are static fracture (15%), stress corrosion cracks (3.5%) and delayed cracks (0.7%) (S. Nishida, 1982, unpublished data) [8,9].

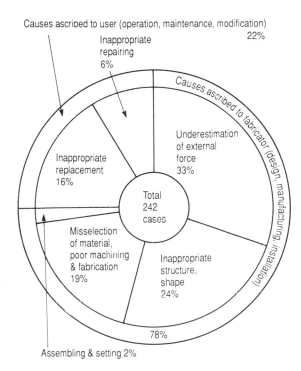

Causes ascribed to user (operation, maintenance, modification) 22%

Inappropriate repairing 6%

Inappropriate replacement 16%

Causes ascribed to fabricator (design, manufacturing, installation)

Underestimation of external force 33%

Total 242 cases

Misselection of material, poor machining & fabrication 19%

Inappropriate structure, shape 24%

78%

Assembling & setting 2%

Figure 1.6 Classification of failures according to factor

Figure 1.6 shows the classification of failures by factors (S. Nishida, 1982, unpublished data) [8,9], which are an underestimation of external force (35%), poor fabrication (22%), inappropriate standards for use and replacement (16%), inappropriate selection of materials (10%), etc. Almost all failures are caused by human factors, and the failures can be prevented by conducting prior analyses, acquiring a good knowledge of materials, and paying attention to inspection after fabrication.

References

1. Frost, N. E. and Dugdale, D. S. (1958) *Journal of the Mechanics and Physics of Solids*, **6**, 92
2. Nishida, S., Urashima, C. and Masumoto, H. (1981) In *The 26th Japan National Symposium on Strength, Fracture and Fatigue*, p. 91, The Japanese Society for Strength and Fracture of Materials, Sendai, Japan
3. Kitsunai, Y. (1970) *Safety Engineering*, **9**, 249, Tokyo
4. Kitsunai, Y. (1974) *Safety Engineering*, **13**, 235, Tokyo
5. Hosino, J. (1973) *Journal of Japan Society of Mechanical Engineers*, **76**, 1126 (including papers by K. Sakamoto, Inoue, Tsumura, F. Terasaki, Y. Nagao, etc.), Tokyo
6. Sakamoto, K. (1975) *Study on Machinery*, **27**, 278
7. Nishida, S. (1983) *Journal of The Japan Welding Society*, **52**, 594
8. Nishida, S. (1982) *Maintenance*, No. 30, 33
9. Nishida, S. (1982) *Maintenance*, No. 31, 40

2   Procedure in failure analysis

2.1 Failure analysis items

When a failure has occurred or a crack is observed in a member of a structure, measures must be taken to prevent similar failures by investigating and analysing the failure. Items necessary for the investigation of failures are given in Table 2.1 (S. Nishida, C. Urashima and H. Masumoto, 1976, 1979 and 1980, unpublished report).

Table 2.1 *Items necessary for investigation of failures*

1. Material used
   (a) Production data: melting, rolling, forming, heat treatment and machining processes
   (b) Chemical analysis: X-ray examination, chemical composition, impurity distribution pattern
   (c) Mechanical properties: tensile, bending, hardness, impact and fatigue tests
   (d) Metallurigical structure: macro- and microstructures of cross section, sulphur print
   (e) Surface treatment and residual stress: finishing
   (f) Fracture surface: external appearance, scanning electron microscope (SEM)
2. Design stress and service conditions
   (a) Assumptions for calculation
   (b) External force: type, magnitude, repetitions
   (c) Atmosphere: air, water, sea-water, etc.
   (d) Others: repair condition, etc.
3. Simulation test
   (a) Laboratory test: stress calculation (strength of materials, finite element method (FEM)), fatigue strength, fracture toughness, stress measuring (strain gauge, photo-elasticity)
   (b) Field confirmation test: stress measurement, production test
4. Overall examination of the results

2.1.1 Investigation of materials used

It is very important to grasp the basic properties of materials in order to obtain basic data for later dynamic calculations, although this may not be directly useful for the calculation of the cause of a failure. In particular, production data, chemical composition, microstructure and mechanical properties are important. In very rare cases, the materials specified in design drawings are not used, or overlay repair welding is carried out during or after fabrication. However, the facts are established by investigating the materials used and this makes it possible to prevent a wrong conclusion in a failure analysis.

Since the fractured part of a member provides a clue to the cause of the fracture, the analysis of fractured surfaces is specially important (see Section 2.4.3). The recent development of the electron microscope permits the cause of a failure to be established from the results of observation under a scanning electron microscope (SEM). Furthermore, the application of techniques of fracture mechanics sometimes makes it possible to calculate stresses applied to a member and their cycle numbers and even to estimate the residual life of an unbroken member.

2.1.2 Investigation of design and service conditions

When machines are designed calculations should be made on the basis of certain preconditions. If a failure occurs in equipment that was designed under conditions which exclude any fracture, this means that the preconditions were wrongly set. By ascertaining whether there is a gap between the preconditions and the actual service conditions, it possible to create new designs by feeding back data. Therefore, it is important to investigate the design and service conditions as minutely as possible. Examples of failures in actual facilities may be utilized as valuable experimental data in this way.

2.1.3 Simulation tests

It is often the case that the truth cannot be established from the results of investigations 2.1.1 and 2.1.2 above alone. In such cases, it is necessary to carry out laboratory tests and to make stress measurements in the field. Simulation tests involve high costs and a long period of time and are therefore impossible to carry out in an emergency.

2.1.4 Overall examination

The main factors are analysed on the basis of the results of steps 2.1.1–2.1.3 above, and the cause of a failure is established. In this instance it is desirable that a conclusion be reached after discussions by specialists instead of being drawn by a single individual. Necessary measures can be thought out by establishing the cause of a failure and by conducting the procedure as quantitatively as possible.

The above procedure is summarized in Figure 2.1.

2.2 Precautions in actual failure analysis

The first thing to do at the sight of a damaged member is to inspect the appearance and fracture surface with the naked eye or by using a loupe and to record the result of the observation as minutely as possible. If the member is immediately cut into pieces, the evidence may be lost. 'From macroscopic observation' is the guiding principle of investigation. When the fracture surface is cut, it is often marked using paint, oil paint, etc. However, such marks on the part to be inspected prevent us from obtaining information from the fracture surface. The fracture surface has irregularities which are microscopically far greater than expected, and these paint and oil paint marks on the fracture surface cannot easily be removed even by cleaning with a solvent.

When a structural member is too hard to cut with a saw, it is advisable to cut it with a grindstone. When the member is large, it is advisable to gas-cut the member in advance. When the electric discharge process is adopted, water glass used as the cooling medium adheres to the fracture surface and covers it. It is therefore necessary to protect this part beforehand when the use of this process is unavoidable. Fracture surfaces of test pieces obtained in the laboratory are unlike almost all actual fracture surfaces of members since rust and other substances adhere to the latter, and the fracture surfaces rub against each other in almost every instance. Methods of removing deposits are described in the next section. When fracture surfaces rub against each other, they should be inspected as a precautionary measure, although the inspection is unlikely to yield any useful information. This is

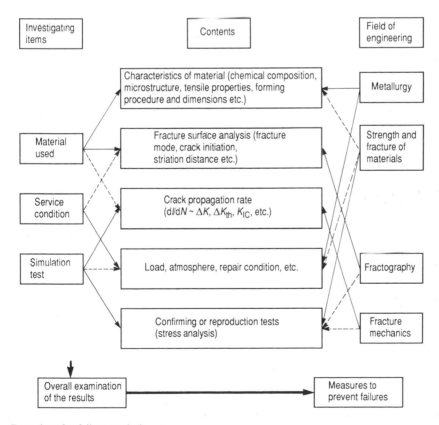

Figure 2.1 Procedure for failure analysis

because, in some cases, parts of the fracture surfaces remain microscopically intact without mutual rubbing. If the fracture surfaces are totally damaged, conclusions must be drawn from information obtained by other means than inspection of the fracture surfaces.

2.3 Methods of removing metal deposits

In a rapidly increasing number of cases the inspection of fracture surfaces is carried out in addition to analysis of failures of structural members and analysis and quantitative evaluation after laboratory tests. In most cases, the fracture surfaces of actual members are covered with deposits such as rust, or they rub against each other. In these cases, the surface deposits (mostly corrosive deposits) should be removed before the fracture surface is inspected.

However, the configuration of the fracture surface, size and shape of cracks, etc. must not be changed when removing the surface deposits. The deposits must be removed in the shortest possible time in such a way that the amount of blank (i.e. melted steel base) is small and the fracture surface is not damaged. Methods of removing iron-based deposits and the results of some experiments are described in the following.

2.3.1 Methods of removing metal deposits

Examples of methods of removing deposits from fracture surfaces are given in Table 2.2 [1–4]. They are broadly classified into

1. The mechanical peel-off method
2. The pickling method, and
3. The cathodic electrolysis method.

In the mechanical peel-off method, an acetyl cellulose film is first immersed in acetone or methyl acetate, and is then attached to the fracture surface in the same procedure as with the transmission electron microscope, and rust is allowed to adhere to the film in order to clean the fracture surface. Although the fracture surface is rarely damaged in this method, the application of the method is limited. In other words, it is almost impossible to remove rust, such as Fe_3O_4, that has good adhesion to the steel base, although rust that has poor adhesion to the steel base, such as Fe_2O_3, can be removed if it is present only to a slight extent. To remove the rust the procedure must be repeated, and skill is required to some extent because a manual operation is involved here. This is therefore quite a troublesome method.

As mentioned above, the configuration of the fracture, size and shape of cracks, etc. must not be changed when removing the surface deposits. In addition, the deposits must be removed in the shortest possible time in such a way that the amount of blank (i.e. melted base steel) is small and the fracture surface is not damaged. In this respect the pickling method is superior to the cathodic electrolysis method. The pickling method is therefore adopted in many cases. From our past experience, the author recommends a pickling method where ultrasonic cleaning is carried out using a 10% H_2SO_4 aqueous solution with 1% inhibitor (Neolecithin). Brand name and manufacturer are: Neolecithin A175, Kitara Kogyo Co. Ltd (Tohsato-onomachi Bld. 501, 1-4-25, Tohsato-onomachi, Sumiyoshi-ku, Osaka 558, Japan, telephone 06-693-8531), pH 8–9, 40% isopropyl alcohol, 15% acetic compound, 10% surface active agent, 1–2% amine-base compound, 30% water, coloured to keep quality, less poisonous, smells like isopropyl alcohol. In this case, an appropriate ultrasonic cleaning time, which varies depending on the amount of deposited rust, is 5 to 15 minutes. Although a solution

Table 2.2   Examples of methods of cleaning fracture surfaces

Classification	Process and contents		
(a) Mechanical peel-off method			
(b) Pickling method	10% H_2SO_4 aqueous solution + inhibitor (Neolecithin 1%) 0.5% ETA* aqueous solution	Immersion or ultrasonic cleaning	
	3% HCl alcohol solution + 2 g/l hexamethylene 6N HCl alcohol solution + 2 g/l hexamethylene tetramine	Immersion for 1 -10 min or ultrasonic cleaning for 5-30 min	
	50% citrate aqueous solution + 50% ammonium citrate aqueous solution (for severe rust) Acetone + 0.5-1% hydrochloric acid (for severe rust)	Immersion or ultrasonic cleaning	
(c) Cathodic electrolysis	1% H2SO4 aqueous solution + inhibitor (IBIT 600), NaOH 500 g + H_2O) (500 cc in total)	Voltage 15 V, current 4 A/cm² for 1-30 sec	

*ETA: Ethylenediaminetetra-acetic acid disodium salt.

at room temperature may be used, it can be raised to 70–80°C for reduction of removing time or when the rust is difficult to remove. The advantage of this method is that the fracture surface is not greatly damaged even if the member is carelessly left immersed in the solution for a long time by an experimenter who may sometimes be engaged in other operations. The rust removal effect is also obtained from a method using a 0.5% ETA (Ethylenediaminetetra-acetic acid disodium salt, Kanto Chemical Co. Ltd) aqueous solution. This method has the advantage of simplicity. However, care should be taken because pitting may occur or the fracture surface may dissolve when the member is immersed for a long time. Nagao [5] also confirmed a similar effect by using a 1–2% ETA aqueous solution, and applied the method to Al brass, low-alloy steels, Tarkalloy cast iron, austenitic stainless steels, Al-alloy steels, titanium, heat-resistant steels, carbon steels, etc. An outline of the results is given in the following.

(a) Al brass: Rust is removed almost completely within 30 minutes. The fracture surface is not damaged.

(b) Low-alloy steels: Quite a large amount of rust remains after 10 minutes and over-etching patterns develop locally within 30 minutes.

(c) Tarkalloy cast iron: The rust removal rate is low. The pearlite of a matrix is etched after more than 30 minutes, with the result that a lamellar structure may appear.

(d) Austenitic stainless steels: The rust removal rate is low. Rust remains to a considerable extent even after 30 minutes. The fracture surface is hardly damaged even after rust removal for more than 30 minutes.

(e) Al-alloy casts: The rust removal rate is high. Rust is removed almost completely within 10 minutes. The fracture surface is not damaged.

(f) Heat-resistant steels: The rust removal rate is high. Rust is removed almost completely within 30 minutes. The fracture surface is not damaged.

(g) Carbon steels: Although the rust removal rate is high, the fracture surface is greatly damaged after more than 30 minutes.

From the above results, Nagao concludes that the higher the corrosion resistance of a material, the higher the rust removal rate and the less the damage to the fracture surface of the material.

Murata and Mukai [6] carried out experiments using a 50% Shumma (250-BC) (Sasaki Chemical Co. Ltd, 1-11, Doshu-machi, Higashi-ku, Osaka, 541, Japan, telephone 03-946-8681) + 50% methyl alcohol solution to remove rust from the fracture surface. Shumma is a reduction-type rust-removing agent comprising 25.0% thioglycolic acid, 25.0% ammonia liquor, 0.1% diethylene triamine sodium acetate, 0.5% oxy-fatty acid alcohol, and 0.1% amid cork acid mono-estel sodium polyvinyl pyrovidon. It seems that this reduction-type rust-removing agent can remove rust without greatly damaging the fracture surface. However, when rust is removed from a rusted fracture surface using this agent, the structure becomes quite corroded and may become etched.

Because there is usually only one fracture surface to be investigated, great care should be taken not to damage the fracture surface during rust removal. In addition, it should be noted that an actual rusted fracture surface is very different from typical surfaces as obtained in the laboratory.

Some results of experiments on pickling methods are described below with reasons why the ultrasonic cleaning method with a 10% H_2SO_4 aqueous solution + 1% inhibitor (Neolecithin) is recommended as a rust removal method.

2.3.2 Methods and results of experiment and discussions

(a) Comparison of pickling methods
The following four methods of removing rust were selected.

1. Immersion in a slightly ammoniacal 20% aqueous solution (80°C), the Japan Petroleum Institute standards for test specimens and holders for the plant test of heat exchangers.
2. Electrolysis at room temperature in a 10% ammonium citrate aqueous solution (cathodic current density 1.1 A/dm^2), the Japan Petroleum Institute standards for test specimens and holders for the plant test of heat exchangers.
3. Immersion in 5% H_2SO_4 + 0.5% Neolecithin aqueous solution (80–90°C), the Japan Petroleum Institute standards for test specimens and holders for the plant test of heat exchangers.
4. Immersion in a 10% H_2SO_4 + 0.5% Neolecithin aqueous solution (80–90°C), the Japan Petroleum Institute standards for test specimens and holders for the plant test of heat exchangers.

For the blank test, the weldable high tensile steel WT80C was used. Five specimens of WT80C, $3.0 \times 10 \times 115$ mm, were finished by plane machining. After oil removal with acetone and ethyl alcohol, the specimens were immersed in the above solution for 1, 3, 5, 10, 20 and 30 minutes. Each specimen was then washed with water while it was being brushed with absorbent cotton. Immediately after that, it was dried with hot air and the loss in mass was measured.

For the test with corrosive deposits, five specimens of WT80C which had the same shape as that used in the blank test were used. A load of $0.6\sigma_y$ was applied to each specimen according to the 4-point bending method and the specimen was immersed in a 500 ppm H_2S solution for seven days and iron sulphide was allowed to generate and become attached to the specimen. After that, rust was removed in the same way as in the blank test and the loss in mass was measured. The results of the measurement are shown in Figures 2.2–2.5, and changes in the surface of each specimen before and after rust removal are shown in Figure 2.6. This photograph shows a comparison of the surface of the specimen before and after each rust removal treatment. In the slightly ammoniacal citrate method, the blank is large and it take 5–10 minutes to remove rust completely. In the ammonium citrate electrolysis method, it takes about 20 minutes to remove rust completely although the blank is small. The greatest disadvantage of this method is that the

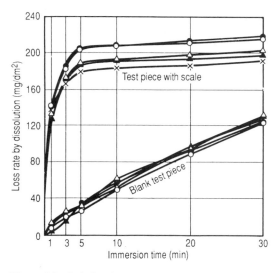

Figure 2.2   Relation between immersion time in a 20% slightly ammoniacal ammonium citrate aqueous solution (80°C) and loss rate by dissolution

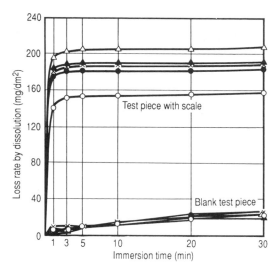

Figure 2.4   Relation between immersion time in a 5% $H_2SO_4 + 0.5\%$ Neolecithin aqueous solution (80–90°C) and loss rate by dissolution

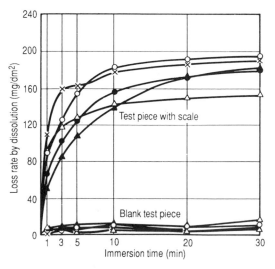

Figure 2.3   Relation between immersion time of 10% ammonium citrate aqueous solution electrolysis (at room temperature) and loss rate by dissolution

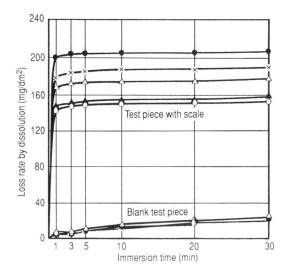

Figure 2.5   Relation between immersion time in a 10% $H_2SO_4 + 0.5\%$ Neolecithin aqueous solution (80–90°C) and loss rate by dissolution

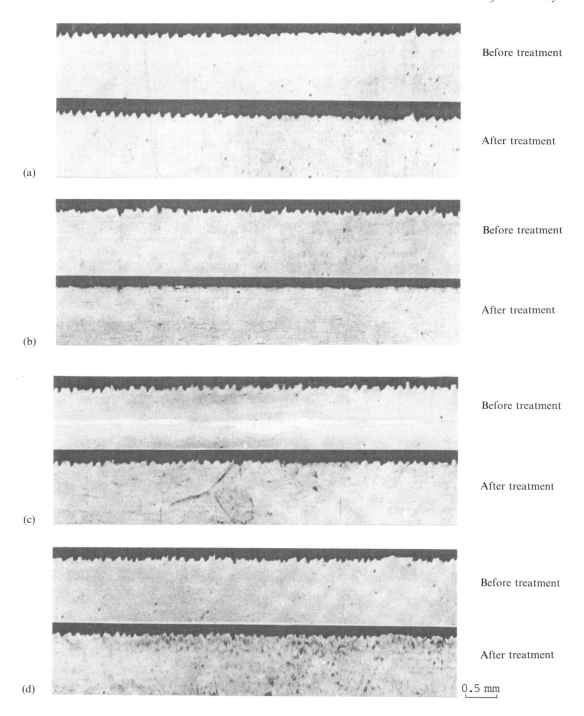

Before treatment

After treatment

(a)

Before treatment

After treatment

(b)

Before treatment

After treatment

(c)

Before treatment

After treatment

(d)

0.5 mm

Figure 2.6 Various rust removal methods and changes in specimen surface for an immersion time of 20 min: (a) 20% slightly ammoniacal ammonium citrate aqueous solution at 78–80°C; (b) 10% ammonium citrate aqueous solution with electrolysis at room temperature and cathodic current density of 1.1 A/dm²; (c) 5% H_2SO_4 + 0.5% Neolecithin aqueous solution at 80–90°C; (d) 10% H_2SO_4 + 0.5% Neolecithin aqueous solution at 80–90°C

surface of the specimen is charged by electrolysis. In the electrolysis method, the irregularities of the surface of the specimen tend to disappear and become flat because the pointed part of the specimen is preferentially melted. In other words, this method cannot be recommended because the specimen surface is damaged.

In the sulphuric acid method, rust is completely removed by immersion for a short time (3 to 5 minutes). The blank is very small and the surface of the specimen is kept in the initial condition even after treatment. There is hardly any difference in the rust removal effect between 5% and 10% sulphuric acid. Because of the small blank, immersion for a long period is allowed and the method is very easy to carry out. It seems that a good result is obtained by adding 1% Neolecithin (inhibitor).

In methods 3 and 4, as will be described later, the surface of the specimen is not greatly damaged even if the specimen is carelessly kept immersed for a long period of time. Moreover, the rust removal effect of these methods is good. Although the solution temperature is kept at 80–90°C, the tests may be carried out at room temperature when ultrasonic cleaning is being used. When rust is very difficult to remove, good results are obtained by keeping the solution temperature at 80–90°C and using ultrasonic cleaning in combination.

(b) Observation of fracture surfaces under SEM with the Neolecithin pickling method

In (a) above, it was found that the Neolecithin pickling method is suitable for rust removal. To demonstrate that an actual fracture surface is not damaged by this method and that an excellent rust removal effect is obtained, some examples of the observation of fracture surfaces under SEM are shown below.

(i) *Method of experiment*

Pickling liquor: 10% sulphuric acid water + 1% Neolecithin solution. Temperature: 25°C (almost constant); ultrasonic cleaning in the pickling solution. Time: 0 (before pickling), 10 and 30 minutes. Fracture surface of specimen: ductile, brittle and fatigue fracture surfaces (all in the atmosphere; laboratory specimens) and corrosion fatigue fracture surface (pH 13, synthetic sea-water + NaOH, dropping rate of 70 cc/min, $N_f = 2.14 \times 10^6$ cycles, 300 rev/min), of the material of low carbon steel.

(ii) *Results of observation under SEM*

Figures 2.7–2.10 show typical examples of ductile, brittle and fatigue fracture surfaces in the atmosphere in the laboratory. The purpose is to show photographs of fracture surfaces that provide basic data for fractography and to demonstrate that the fractures are

little damaged by rust removal from the surface, which is necessary for actual fracture analysis. Figure 2.7 shows a ductile fracture surface. Inclusions are observed at the bottom of a somewhat large dimple in the fracture surface before cleaning. Some of these inclusions disappear after ultrasonic cleaning for 10 minutes in a 10% sulphuric acid water + 1% Neolecithin solution. The remaining inclusions disappear completely after ultrasonic cleaning for 30 minutes. It is found, however, that the configuration of the fracture surface hardly changes even after ultrasonic cleaning for 30 minutes. Figure 2.8 shows brittle fracture surfaces. Figure 2.9 shows enlarged photographs of the fracture surfaces of Figure 2.8. The fracture surface shown in Figure 2.8(c) was further immersed in the cleaning liquid at 85°C for 30 minutes. The fracture surface shown in Figure 2.8(e) was obtained when that shown in Figure 2.8(d) was subjected to ultrasonic cleaning for 3 minutes after immersion in synthetic sea-water for 10 days. It is apparent from these photographs that although the fracture surface is hardly damaged even if it is subjected to ultrasonic cleaning in a 10% sulphuric acid water + 1% Neolecithin solution or if it is immersed in a heated-up solution, it may be greatly damaged by cleaning if it is covered with rust.

Figure 2.10 shows fatigue fracture surfaces. Figure 2.11 shows enlarged photographs of the fracture surfaces shown in Figure 2.10. The foregoing applies to each of the photographs. Incidentally, striations are observed in Figures 2.10 and 2.11; they are peculiar to fatigue fracture surfaces and their intervals are about 1 μm. These intervals correspond almost to the crack propagation rate per cycle.

Figure 2.12 shows a case where ultrasonic cleaning (for 30 minutes) with a 10% sulphuric acid water + 1% Neolecithin solution was conducted for a corrosion fatigue fracture surface that was obtained under the conditions: stress amplitude σ_a of 20 kgf/mm^2, number of cycles to failure N_f of 2.14×10^4, pH 13 with synthetic sea-water + NaOH, drip rate of 70 cc/min, and cycle ratio of 300 rev/min. The top portion of each photograph shows the surface near the initiation point of fracture. Before cleaning, rust adheres and the configuration of the fracture surface is scarcely identified. Although almost all the rust is removed by cleaning (Figures 2.12(b) and (d)), the configuration of the fracture surface cannot yet be easily recognized from the photographs alone. This may be because the area near the initiation point of fracture is shown and because the original shape of the fracture surface was greatly changed by the adhering rust. It is known beforehand that this fracture surface is a fatigue fracture made in the laboratory at the initiation point of fracture. Therefore, this fracture surface can be judged to be a fracture surface even if striations peculiar to fatigue are not observed. In any case, almost all actual members are subjected to a corrosive

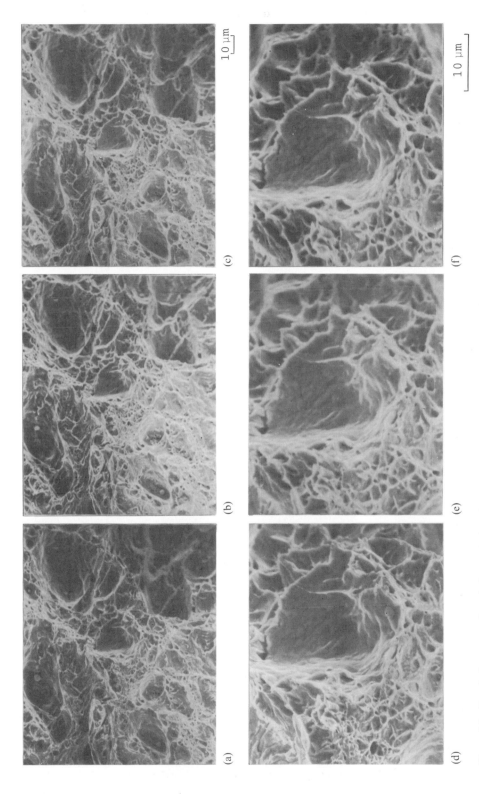

Figure 2.7 Ductile fracture surfaces before and after ultrasonic cleaning in 10% sulphuric acid water + 1% Neolecithin at 25°C: (a) and (d) before cleaning; (b) and (e) after cleaning for 10 min; (c) and (f) after cleaning for 30 min ((d), (e) and (f) are enlargements of (a), (b) and (c), all taken at the same point)

Direction of crack propagation

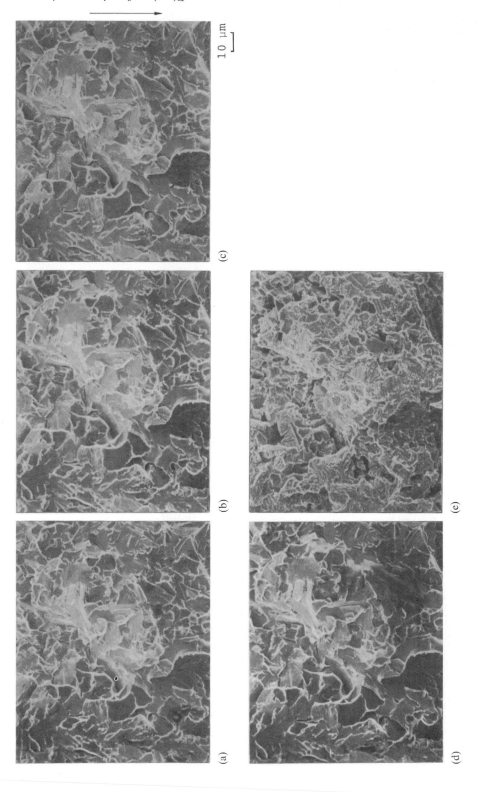

Figure 2.8 Brittle fracture surfaces before and after ultrasonic cleaning in 10% sulphuric acid water + 1% Neolecithin at 25°C: (a) before cleaning; (b) after cleaning for 10 min; (c) after cleaning for 30 min; (d) following treatment (c) and further immersion at 85°C for 30 min; (e) following treatment (d) and immersion in synthetic sea-water for 10 days and ultrasonic cleaning for 3 min (all taken at the same point)

Direction of crack propagation

10 μm

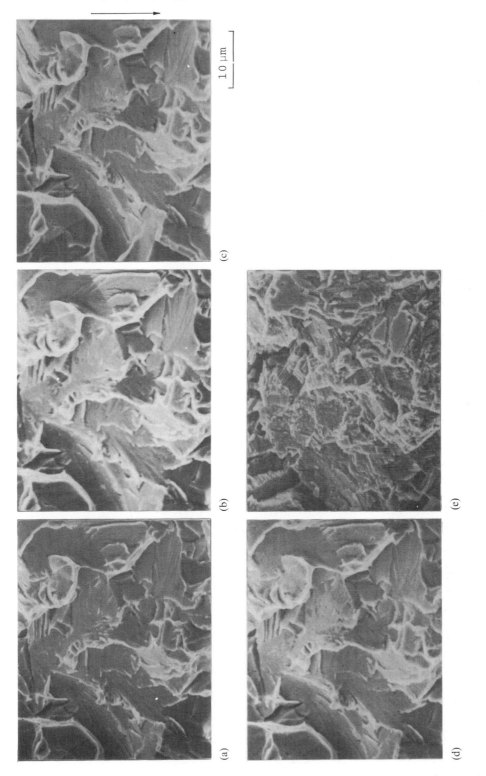

(a)   (b)   (c)   (d)   (e)

Figure 2.9   Brittle fracture surfaces (enlargements of those in Figure 2.8)

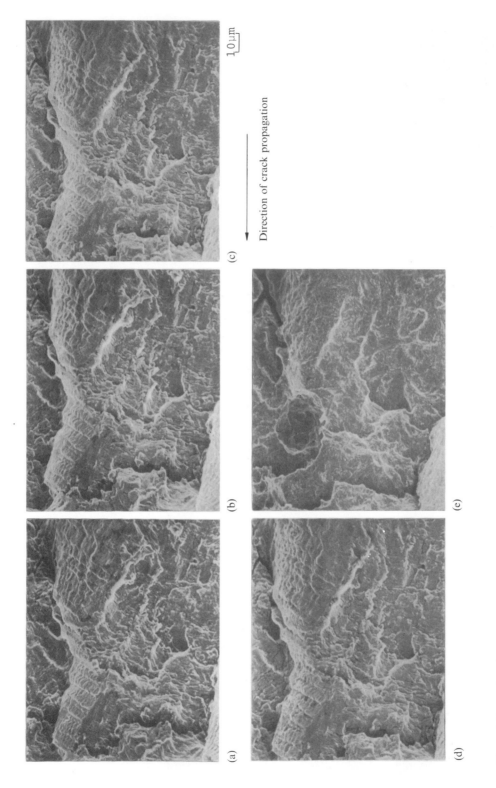

Direction of crack propagation

10 μm

(a)

(b)

(c)

(d)

(e)

Figure 2.10 Fatigue fracture surfaces (details as for Figure 2.8)

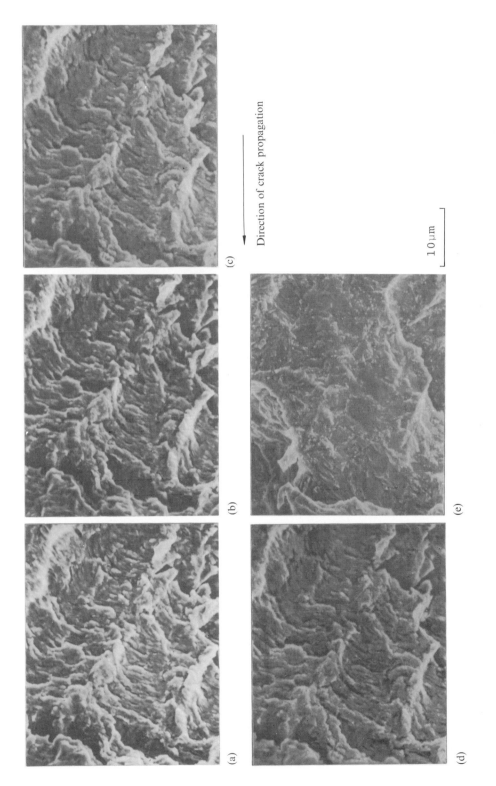

Direction of crack propagation

10 μm

(a) (b) (c) (d) (e)

Figure 2.11 Fatigue fracture surfaces (enlargements of those in Figure 2.10)

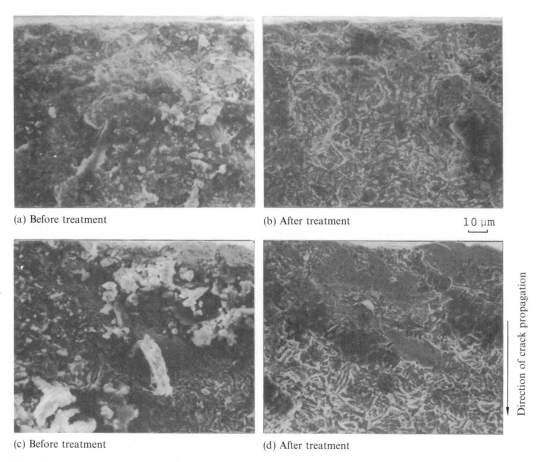

(a) Before treatment   (b) After treatment   10 μm

(c) Before treatment   (d) After treatment

Direction of crack propagation

Figure 2.12   Ultrasonic cleaning (for 30 min) of corrosion fatigue specimen in 10% sulphuric acid water +1% Neolecithin at 25°C. Specimen was obtained under the conditions: stress amplitude, $\sigma_a = 20$ kgf/mm^2; number of cycles, $N = 2.14 \times 10^4$; frequency, 300 rev/min; and drip rate of synthetic sea-water + NaOH solution, 70 cc/min (all taken at the crack initiation point of the fracture surface)

environment. It will be understood therefore that it is not easy to carry out appropriate failure analyses because actual fracture surfaces are different from examples of failure surfaces obtained in the laboratory.

References

1. Nishida, S. and Mizoguchi, S. (1985) The Society of Materials Science (JSMS), Committee on Fractography, Subcommittee on Environmental Fracture, Preprint FE16, 1 (in Japanese), Kyoto, Japan
2. Fujiwara, M. (1980) *Journal of the Society of Materials Science*, **29**, 1247, Kyoto, Japan
3. Nishida, S. and Mizoguchi, S. (1984) Sectional Meeting on Environmental Strength of Steels, Preprint ISE-6-5, 1, Tokyo
4. Ebara, R. (1984) Sectional Meeting on Environmental Strength of Steels, Preprint ISE-6-5, 1, Tokyo
5. Nagao, Y. (1985) The Society of Materials Science (JSMS), Committee on Fractography, Subcommittee on Environmental Fracture, Preprint FE15, 1, Kyoto, Japan
6. Murata, M. and Mukai, Y. (1985) The Society of Materials Science (JSMS), Committee on Fractography, Subcommittee on Environmental Fracture, Preprint FE18, 1

2.4 Useful literature on failure analysis

Chapter 2 has described the procedure for failure analysis, but the description is not exhaustive because of limited space. The following literature, mainly books, may be useful for failure analysis.

(a) Literature on fractography

1. ASM Handbook Committee (1968) *The ASM Metals Handbook – Vol. 4: Testing, Metallography and Failure Analysis*, ASM
2. Koterasawa, R. (1974) *Journal of the Society of Materials Science* (JSMS), **23**, No. 248, 412; No. 249, 479; No. 250, 593; No. 251, 666; No. 252, 803 (in Japanese)
3. McDonnel Douglas Astronautics Company (1975) *SEM/TEM Fractography Handbook*, Metals and Ceramics Information Center, Battelle
4. Philips, A., Kerlins, V., Rawe, R. A. and Whiteson, B. V. (1976) *Electron Fractography Handbook*, Metals and Ceramics Information Center, Battelle
5. Kanto Branch of the Society of the Electron Microscope (ed.) (1976) *Scanning Electron Microscope – Its Basis and Application*, Kyoritsu-Shuppan, Tokyo
6. Kitagawa, H. and Koterasawa, R. (1977) *Fractography*, Baifukan, Tokyo
7. Koterasawa, R. (1978) *Journal of the Society of Materials Science* (JSMS), **27**, No. 300, 871 (in Japanese)
8. Koterasawa, R. (1979) Teaching Material for the Kansai Branch of the Japan Society of Mechanical Engineers, 13
9. Matsuda, F. and Nakagawa, H. (1979) *Collection of Fractography on Welds*, Kuroki-Shuppan Co. Ltd, Tokyo
10. Bhattacharyya, S., Johnson, V. E., Agarwal, S. and Howes, M. A. H. (1979) *Failure Analysis of Metallic Materials by Scanning Electron Microscopy*, The Metals Research Division, IIT Research Institute
11. The Society for Studying Screw Loosening Fractures (1980) *Loosened Fracture of Screws and Bolts*, Management and Development Center, Tokyo
12. The Society for Studying Measures against Crack and Fracture of Metals (1980) *Crack and Fracture of Metals – Its Analysis and Countermeasures*, Management and Development Center, Tokyo
13. Koterasawa, R. (1981) *Fractography and Its Application*, Nikkan Kogyo Newspaper Co. Ltd, Tokyo
14. Committee for Studying Welds and Metallurgy of the Japan Welding Society (1982) *Collection of Photographs of Fracture Surfaces of Welds of Irons and Steels*, Kuroki-Shuppan, Tokyo
15. Koterasawa, R. (ed.) (1985) *Collection of Photographs of Fracture Surfaces of Steels*, Techno-eye, Tokyo
16. Committee of Environmental Strength of Irons and Steels (1985) *Collection of Photographs of Marine Environmental Fracture Surfaces of Iron and Steel*, Vol. 1

(b) Literature on fatigue data and fracture toughness

1. The Japan Society of Mechanical Engineers (JSME) (1961) *Design Data on Fatigue Strength of Metals*, Vol. 1
2. The Japan Society of Mechanical Engineers (JSME) (1965) *Design Data on Fatigue Strength of Metals*, Vol. 2
3. The Society of Materials Science (SMS) (1968) *Handbook of Fatigue Testing of Metals*, Yokendo, Tokyo
4. Royal Swedish Academy of Engineering Science (1969) Monograph on Fatigue Strength of Welds, Sections 1 and 2
5. Committee of Gakushin No. 129 (1970) *Collection of Strength and Fatigue Data of Metals*, Maruzen, Tokyo
6. Nippon Steel Corporation (1970) Technical Report, Fatigue Data Sheet, Vol. 1
7. Nippon Steel Corporation (1971) Technical Report, Fatigue Data Sheet, Vol. 2
8. The Society of Materials Science (SMS) (1973) *Design Handbook for Fatigue of Metals*, Yokendo, Tokyo
9. The Society of Materials Science (SMS) (1982) *Collection of Fatigue Strength Data of Metals*, Vols 1, 2 and 3, Kyoto, Japan
10. The Iron and Steel Institute (1982) *Fracture Toughness of Structural Steels*
11. The Society of Materials Science (SMS) (1983) *Collection of Fatigue Crack Propagation Rate Data of Metals*, Vols 1, 2 and 3
12. Committee on Environmental Strength of Iron and Steel (1983) *Data Base of Marine Environmental Strength of Irons and Steels*, Vol. 1, Tokyo
13. Committee on Environmental Strength of Iron and Steel (1985) Preprints of Second Symposium, Tokyo

(c) Literature on failure analysis and examples of failure

1. Yoshida, T. (1974) *How to Observe the Fracture Surface of Metals*, Nikkan Kogyo Newspaper Co. Ltd, Tokyo
2. Shimoda, H. (1975) *Investigation and Analysing Method and Study of Causes of Failures of Iron and Steel Products*, Ohyogijutsu-Shuppan, Tokyo
3. Henry, G. and Hortmann, D. (1979) *De Ferri Metallographia*, Verlag Stahleisen mbH, Dusseldorf
4. Nagaoka, K. (1979) *Failure Analysis of Machine Parts*, Kogaku-tosho, Tokyo
5. The Japan Society of Mechanical Engineers (JSME) (1984) *Case Study and Analysis Technology of Failures of Machines and Structures*
6. Von Friedrich, Karl Navman (1985) (translated by U. Hashimoto, T. Yokobori and J. Tsuji) *Case*

Study of Machine Parts and Iron and Steel Materials, Maruzen

(d) Literature on stress intensity factors

1. Sih, G.C. (1973) *Handbook of Stress-Intensity Factors*; Leihigh University
2. Tada, H., Paris, Paul C. and Irwin, George R. (1973) *The Stress Analysis of Cracks Handbook*, Del Research Corporation
3. Roobe, D.P. and Cartwright, D.J. (1976) *Stress Intensity Factors*, Her Majesty's Stationery Office, London
4. Committee of Fracture Mechanics of SMS (1986) *Stress Intensity Handbook*, Pergamon Press, Oxford

(e) Literature on material strength, especially fatigue

1. Kawamoto, M. (1962) *Fatigue of Metals*, Asakura-shoten, Tokyo
2. Ishibashi, T. (1964) *Strength of Metals*, Yokendo, Tokyo
3. The Japan Institute of Metals (JIM) (1964) *Strength and Fracture of Metallic Materials*, Maruzen, Tokyo
4. Yokobori, T. (1964) *Strength of Materials*, Gihodo, Tokyo
5. The Society of Materials Science (SMS) (1964) *Fatigue of Metals*, Maruzen, Tokyo
6. Kawata, Y. (1966) *Fatigue of Metals and Design*, Ohm-sha, Tokyo
7. Knott, J.F. (1966) (translated by H. Miyamoto) *Basis of Fracture Mechanics*, Baifukan, Tokyo
8. Ishibashi, T. (1967) *Fatigue of Metals and Prevention of Fracture*, Yokendo, Tokyo
9. Ishibashi, T. (1967) Collection of Dr T. Ishibashi's Papers, Yokendo, Tokyo
10. Yokobori, T. (1968) *Strength of Materials*, Iwanami-shoten, Tokyo
11. Ivanova, V.S. (1970) (translated by T. Yokobori *et al.*) *Fatigue Fracture of Metals*, Maruzen, Tokyo
12. Tetelman, A.S. and McEvily, Jr A.J. (1971) (translated by H. Miyamoto) *Strength and Fracture of Structural Materials*, Vols 1 and 2, Baifukan, Tokyo
13. Kawamoto, M. *et al.* (1972) *Fatigue of Metals and Design*, Coronasha, Tokyo
14. Miyamoto, H. (1972) *Dynamics of Fracture*, Coronasha, Tokyo
15. Cazaud, R. *et al.* (1973) (translated by H. Funakubo and S. Nishijima) *Fatigue of Metals*, Maruzen, Tokyo
16. Forsyth, P.J.E. (1975) (translated by H. Nakazawa, H. Kobayashi) *Basis of Fatigue of Metals*, Yokendo, Tokyo
17. Nakamura, H. and Tanaka, S. (1976) *Calculation Method of Fatigue Life of Machine*, Yokendo, Tokyo
18. Nakazawa, H. and Kobayashi, H. (1976) *Strength of Solids*, Kyoritsu-Shuppan, Tokyo
19. Okamura, H. (1976) *Introduction to Linear Fracture Mechanics*, Baifukan, Tokyo
20. Ishida, M. (1976) *Elasticity of Crack and Stress Intensity Factor*, Baifukan, Tokyo
21. Mura, T. and Mori, T. (1976) *Micromechanics*, Baifukan, Tokyo
22. Ohnami, M. and Shiozawa (1976) *Strength and Fracture of Polycrystals*, Baifukan, Tokyo
23. Kuroki, G. and Ohmori, M. (1977) *Strength and Fracture of Metals*, Morikita-Shuppan, Tokyo
24. Sih, G.C. (1977) *Application of Fracture Mechanics*, Morikata-Shuppan, Tokyo
25. Kanazawa, T. and Koshiga, F. (1977) *Brittle Fracture*, Vol. 2, Baifukan, Tokyo
26. Kachanov, L.M. (1977) (translated by Y. Ohhashi) *Basis of Fracture*, Morikata-Shuppan, Tokyo
27. Koterasawa, R. (1979) *Summary of Strength of Materials*, McGraw-Hill
28. Rorumu, S.T. and Barsom J.M. (1981) (translated by T. Yokobori, T. Kawasaki and J. Watanabe) *Prevention of Fatigue and Fracture of Structures*, Baifukan, Tokyo
29. Endo, K. and Komai, K. (1982) *Corrosion Fatigue of Metals and Design Based on Strength*, Yokendo, Tokyo
30. Nakazawa, H. and Honma, H. (1982) *Fatigue Strength of Metals*, Yokendo, Tokyo
31. Nakamura, H. Tsunenari, Y., Horikawa, T. and Okazaki (1983) *Machine Design on Fatigue Life*, Yokendo, Tokyo
32. Davidson, D.L. and Suresh, S. (1983) *Fatigue Crack Growth Threshold Concepts*, AIME
33. Ohnami, M. (ed.) (1984) *Introduction of Strength of Materials*, Ohmu-sha, Tokyo
34. Tamura, I. and Horiuchi, R. (eds) (1984) *Strength and Physical Properties of Materials*, Ohmu-sha, Tokyo
35. Abe, H. (ed.) (1984) *Analysis of Strength* Vol. 2, Ohmu-sha, Tokyo
36. Ohtani, R. and Komai, K. (eds) (1984) *Environmental and High-Temperature Strength*, Ohmu-sha, Tokyo
37. Asada, Y. and Koibuchi, K. (1984) *Strength of Machine and Structure*, Ohmu-sha, Tokyo
38. Klesnil, M. and Lukas, P. (1984) (translated by T. Araki and S. Horibe) *Fatigue Dynamics and Structure*
39. Okamura, H. (ed.) (1985) *Analysis of Strength*, Vol. 1, Ohmu-sha, Tokyo
40. Ohji, K. (ed.) (1985) *Strength of Fracture*, Ohmu-sha, Tokyo
41. Nisitani, H. (ed.) (1985) *Fatigue Strength*, Ohmu-sha, Tokyo

(f) Literature on machine design
1. The Japan Society of Mechanical Engineers (JSME) (1934) *Mechanical Engineering Handbook*, 1st edn, Maruzen, Tokyo
2. The Japan Society of Mechanical Engineers (JSME) (1968) *Mechanical Engineering Handbook*, 5th edn, Maruzen, Tokyo
3. Kawata, Y. Kawamoto, M., Yokobori, T. and Miyagawa, M. (1972) *Engineering Handbook of Material Strength*, Asakura-shoten, Tokyo
4. Editing Committee for Machine Design Handbook (1973) *Machine Design Handbook*, JSME
5. Editing Committee for Strength Design Data Book (1977) *Strength Design Data Book*, 8th edn, Shokabo, Tokyo
6. Celensen, S. V., Kogaev, V. P. and Schneiderovich, R. M. (tanslated by Ohhashi, Y.) (1978) *Calculation Handbook of Strength of Machine Elements*, Morikata-Shuppan, Tokyo

(g) Literature on SI units and actual stresses
1. Lattice (1965) *Unit Dictionary*, Lattice
2. Endo, T., Mitsunaga, K., Takahashi, K. *et al.* (1974) *Proceedings of the 1974 Symposium on the Mechanical Behavior of Materials*, **1**, 371, ASMS
3. Editing Committee for International Unit System SI Manual (1976) *International Unit System SI Manual*, Japanese Standards Association (JSA)
4. Japanese Standards Association (JSA) (1977) *ISO Standards Table*, JSA
5. The Japan Society of Mechanical Engineers (JSME) (1979) *SI Manual in Mechanical Engineering*, JSME
6. Endo, T. (1980) *Measuring Technology of Fatigue Life for Machinery Structures according to the Rainflow Method*, Fuji-Techno-System
7. Yamada, K. (1984) Fatigue Strength Committee of Welds in JWS, FS-640–84, 1
8. Yamada, K. (1985) *Proceedings of the Japan Society of Civil Engineering and Construction*, **2**, 25
9. See references 17, 19 and 36 in (e)

3   Initiation and propagation of fatigue cracks

As described earlier, it is known that more than 85–90% of failures of structural members are caused by fatigue. Therefore, this chapter describes the mechanism of fatigue failure and some explanations useful for life prediction for structural members with fatigue cracks.

When repeated stresses of sufficient magnitude are applied to a member with no crack or crack-like defect, a crack is formed on the surface after a certain number of cycles. It is well known that this crack propagates when further stresses are repeatedly applied. Because these two phenomena are considered to be different physical phenomena, they are separately discussed from a practical aspect.

3.1   Characteristics of fatigue failure

Fatigue failure occurs when repeated stresses are applied to a member; the failure has the following characteristics:

1. The failure may occur even if the amplitude of repeated stresses is smaller than the static tensile strength or yield strength.
2. The fracture propagates seemingly without plastic deformation. The fracture surface is very smooth although it is usually stepped.
3. When a definite stress amplitude is applied, fracture occurs after a definite number of cycles.
4. An almost linear relation in terms of logarithmic representation exists between the stress amplitude S and the number of fractures to failure N. This relationship is expressed by a curve generally called the S–N curve or the Wöhler curve, after the name of the discoverer (see Figure 3.1).
5. In a carbon steel, a value called the fatigue limit or endurance limit exists. Fracture does not occur when repeated stresses below this stress amplitude

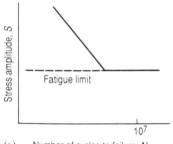

(a)   Number of cycles to failure, N

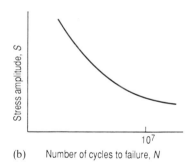

(b)   Number of cycles to failure, N

Figure 3.1   Representative S–N curves: (a) where a distinct fatigue limit exists (in a carbon steel); (b) where a distinct fatigue limit does not exist (in copper, aluminium, etc.)

are infinitely applied. However, the number of cycles to failure tends to fluctuate when stresses above the endurance limit are repeated.
6. The fatigue limit is little affected by the mean stress when the mean stress is lower than the elastic limit. However, when the mean stress is large, to a certain degree, the fatigue limit is affected by the mean stress and is low.
7. Even if stresses below the fatigue limit are repeated, a hysteresis loop can be observed.

3.2 Initiation of fatigue cracks

Fatigue cracks may sometimes be produced even when the stress amplitude is lower than the static tensile strength or the yield strength. Although this will be understood in the case where the amplitude of repeated stress is equivalent to the static tensile stress, it may be difficult to understand the reason why fracture occurs when the amplitude of repeated stress is lower than the yield strength. The reason is explained using a model in the following.

Figure 3.2 shows the difference between the macroscopic yield point and the start of microscopic plastic deformation. Figure 3.2(a) shows a stress–strain curve obtained when uniaxial tension is applied to a typical low-carbon steel specimen. Points S and R represent the upper and lower yield points respectively, and point B represents the tensile strength. The yield point is commonly judged to be the point at which plastic deformation starts in the specimen. In other words, it

might be thought that plastic deformation scarcely occurs below the yield point. However, when the electro-polished surface of an annealed S10C specimen is observed [1], it is found that some of the grains begin to deform plastically at a stress far lower than the yield point (if the yield point is denoted by σ_y, then the stress is about $\sigma_y/3$ (see Figure 3.2(b)). (If a specimen of carbon steel, etc. is electro-polished, the surface structure can be directly observed under an optical microscope. An electrolyte composed of 1000 gf of phosphoric acid, 20 gf of gelatin and 20 gf of oxalic acid is suitable for this purpose.) Because plastic deformation at this stage occurs in a very small portion of the whole volume, residual strains are scarcely detectable after unloading. In any case, it should be noted that there is a difference in stress levels for macroscopic and microscopic deformation.

Figure 3.3 shows the difference between the static tensile test and the fatigue test with the aid of a model. When the load is increased gradually from zero, plastic deformation begins, usually at the location of the specimen surface where slip occurs most easily. If these portions are indicated by the dotted lines in Figure 3.3(a), they deform by the effect of tension as shown in Figure 3.3(b). When the tensile load is reduced from its maximum level and a compressive load is applied, deformation occurs as shown in Figure 3.3(c) if slip

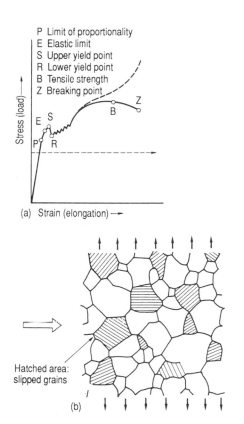

P Limit of proportionality
E Elastic limit
S Upper yield point
R Lower yield point
B Tensile strength
Z Breaking point

(a) Strain (elongation) →

Hatched area: slipped grains

(b)

Figure 3.2 Difference between macroscopic and microscopic deformation: (a) stress–strain diagram; (b) slip grains under stress lower than yield stress (microscopic yield). Yield point (S, R): plastic flow appears in the whole section

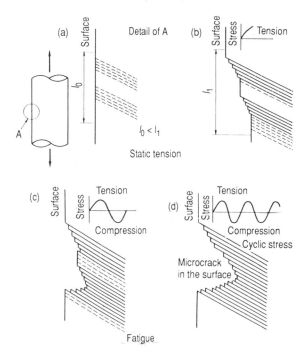

Figure 3.3 A model of the difference in mechanism between static tensile tests ((a) and (b)) and fatigue tests ((c) and (d))

occurs in the portion indicated by the dotted lines in Figure 3.3(b). For the fatigue test the stress is repeated and parts therefore become as shown in Figure 3.3(d). A crack is seen when the specimen is observed from the surface. Once a crack is initiated in this way, stress concentration occurs at both ends of the crack and the crack continues to grow, except in special instances.

It is known that the fatigue limit is the limit stress under which a microcrack is initiated but does not propagate further, and not the limit stress under which a crack is not initiated. The design process is usually conducted in such a way that the stress applied to a member is below the fatigue limit. For literature on design data bases dealing with such a case, see Section 2.4(b).

3.3 Propagation of fatigue cracks

The fatigue crack initiated in the preceding section begins to propagate by the effect of the successive repetition of stress. Because fatigue fracture is often initiated from a notch or crack-like defect in an actual member, importance has been attached to the evaluation of fatigue crack propagation characteristics in the process of service life prediction and safety (see Appendix 2).

Figure 3.4 shows a model of the fatigue crack growth process [2]. Cracks are initiated from a defect-free surface, fabrication defects (including machining flaws, weld defects, and inclusions), notches (including key grooves, and 'step and hole'), etc. In any case, it is apparent that the process of crack growth occupies the greater part of the life of a member. It is known that there is a linear relationship between the crack propagation rate dl/dN and the range ΔK_{eff} of the stress

intensity factor after crack initiation until just before final fracture (see Section 2.4(b)). We evaluated the fatigue crack propagation characteristics of typical steel grades [3], and the results are shown below.

3.3.1 Method of experiment

Mild steel (SS41) and plain carbon steels for machine structural use (S25C–S55C, four kinds), weldable high-tensile steels (SM50–HT80, three kinds), structural low-alloy steels (SCM440 as rolled, and quenched and tempered), and an austenitic stainless steel (SUS304) – a total of 11 steel grades – were used in the test. The chemical compositions and mechanical properties of these materials are given in Table 3.1. The microstructures of the materials are shown in Figure 3.5. The tensile strength ranges from the 40–90 kgf/mm² class and the structures can be broadly divided into ferrite plus pearlite, pearlite, tempered martensite, and austenite. Table 3.1 also shows the rotary bending fatigue limit of specimens taken axially in the rolling direction with their centre corresponding to the 1/4 thickness, and the critical crack opening displacement (COD) value at room temperature, δ_c [4]. COD test specimens (25 mm thick × 50 mm wide × 250 mm long) were also taken in the rolling direction and were tested at room temperature in accordance with ASTM E399.

Figure 3.6 shows the shape of a specimen for the fatigue crack propagation test. Specimens for this test were all taken in the rolling direction from the middle of the thickness except SCM440 (from 1/4 thickness). The measuring part of the crack length was polished axially with no. 600 emery paper.

An electro-hydraulic servo fatigue testing machine of ± 40 tf was used in the test. Two or three specimens per steel grade were tested at a cyclic speed of 900 cycles/minute and a stress ratio R of 0. The crack length was measured by observing both sides of each specimen with a reading microscope of magnification × 20, and measured values were expressed in the average value.

2.3.2 Results of experiment

Figure 3.7 shows the relationship between the fatigue crack propagation rate dl/dN and the range of stress intensity factors of the eleven steel grades used. The stress intensity factor K was calculated from the following equation:

$$K_I = \sigma_a \sqrt{[\pi l \ sec \ (\pi l/W)]} \tag{3.1}$$

where σ_a is the repeated stress, l is the half length of the crack and W is the half width of the specimen.

S35C shows the lowest crack propagation rate, and SUS304 the highest. However, the difference between the highest and lowest values is about 2.5 times at most. The linear relation shown in Figure 3.7 is

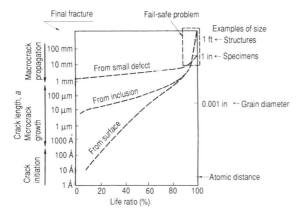

Figure 3.4   A concept of crack growth and stress intensity factor ΔK

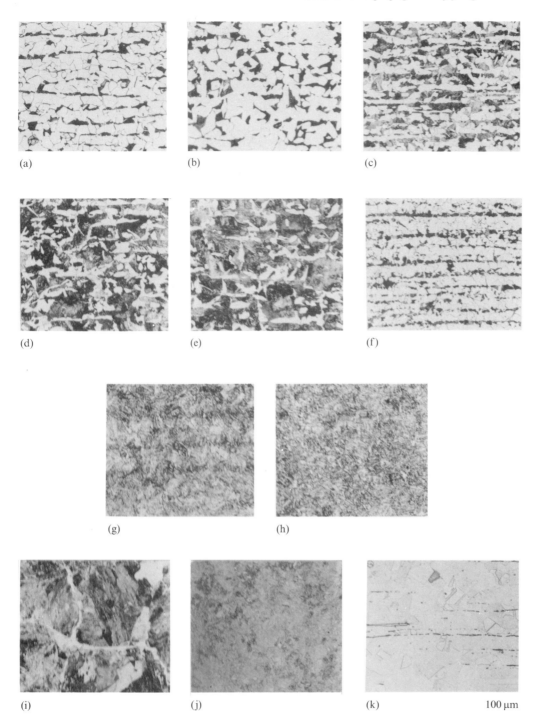

Figure 3.5 Microstructures of materials used: (a) SS41; (b) S25C; (c) S35C; (d) S45C; (e) S55C; (f) SM50; (g) HT60; (h) HT80; (i) SCM440; (j) SCM440; (k) SUS304

Table 3.1 *Chemical composition and mechanical properties of materials used*

Kinds of steel	t (mm)	CD	Chemical composition (wt%)								Mechanical properties					
			C	Si	Mn	P	S	Ni	Cr	Mo	σ_y (kgf/ mm^2)	σ_B (kgf/ mm^2)	El (%)	RA (%)	σ_{wb}(kgf/ mm^2)	COD, δ_c (mm)
SS41	32	L	0.17	0.19	0.63	0.025	0.025	–	–	–	28.5	45.2	37.7	65.1	25.0	1.889
S25C	32	L	0.30	0.23	0.47	0.018	0.018	–	–	–	26.3	49.4	35.9	57.8	21.0	0.958
S35C	32	L	0.42	0.21	0.69	0.022	0.004	–	–	–	34.5	62.1	35.0	53.0	24.0	0.349
S45C	32	L	0.49	0.23	0.77	0.022	0.007	–	–	–	34.3	71.1	22.2	40.9	27.0	0.113
S55C	32	L	0.52	0.18	0.70	0.014	0.014	–	–	–	36.7	71.3	20.2	41.2	25.0	0.066
SM50	35	L	0.18	0.43	1.31	0.025	0.009	–	–	–	35.7	56.0	39.0	72.6	29.0	1.782
HT60	40	L	0.11	0.23	1.25	0.014	0.003	0.16	–	0.13	58.5	69.2	28.6	76.9	35.0	1.213
HT80	36	L	0.11	0.23	0.87	0.013	0.005	0.79	0.50	0.44	71.2	80.6	27.0	74.5	42.0	0.464
SCM440 (as rolled)	131	L	0.42	0.25	0.74	0.020	0.009	0.01	1.08	0.16	57.2	88.1	18.3	35.8	32.0	0.030
SCM440 (QT)	131	L	0.42	0.25	0.74	0.020	0.009	0.01	1.08	0.16	72.3	95.1	22.7	65.7	48.0	0.586
SUS304	35	L	0.06	0.66	1.03	0.028	0.004	9.37	18.95	–	27.1	62.7	62.6	72.1	31.0	–

t, thickness of plate; CD, cut-out direction; L, longitudinal direction; σ_y, yield stress; σ_B, tensile strength; σ_{wb}, fatigue limit of rotating bending; δ_c, critical COD value.

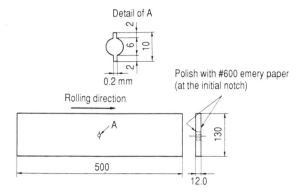

Figure 3.6 Dimensions of the specimen for fatigue crack propagation

Table 3.2 *Values of material constants* C *and* m

Kinds of steel	Symbols	C	m
SS41	1	2.12×10^{-12}	3.89
S25C	2	2.12×10^{-12}	3.89
S35C	3	2.85×10^{-12}	3.70
S45C	4	4.99×10^{-11}	3.20
S55C	5	4.33×10^{-11}	3.24
SM50	M5	2.51×10^{-11}	3.41
HT60	W6	1.71×10^{-10}	2.88
HT80	W8	2.63×10^{-10}	2.82
SCM440 (as rolled)	M 4–25 26	5.20×10^{-9}	2.12
SCM440 (QT)	M 4–27 28	8.71×10^{-10}	2.56
SUS304	SU	8.04×10^{-11}	3.18

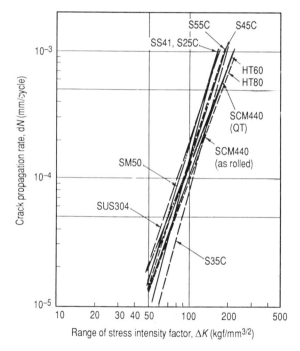

Figure 3.7 Relation between crack propagation rate d*l*/d*N* and range of stress intensity factor Δ*K*

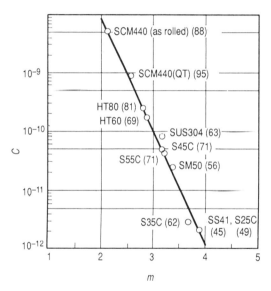

Figure 3.8 Relation between material constants *C* and *m*

rearranged in the following equation and the coefficient is found by the method of least squares.

$$\mathrm{d}l/\mathrm{d}N = C(\Delta K)^m \qquad (3.2)$$

C and *m* are assumed to be material constants and the figures given in Table 3.2 can be obtained. There is a relatively good correlation between these material constants *C* and *m* and the mechanical property σ_B.

Figure 3.8 shows the relationship between the material constants *C* and *m*. There is an almost linear

relationship between *C* and *m* and the following equation can be obtained:

$$\log \frac{1}{\sqrt{C}} = m + 2 \qquad (3.3)$$

The values in parentheses show the tensile strength of each material. The higher the tensile strength, the higher the value of *C* and the lower the value of *m*.

Figure 3.9 shows the relationship between the material constant *C* and the tensile strength. Although the scatter is quite marked, the higher the tensile strength, the higher the material constant *C*.

Figure 3.10 shows the relationship between the material constant *m* and the tensile strength. The same tendency seen in Figure 3.9 is observed. However, there is scarcely any correlation between the material

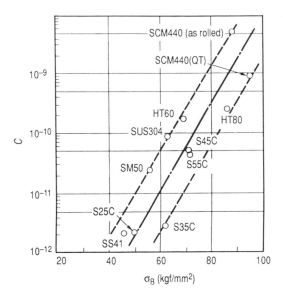

Figure 3.9 Relation between material constant C and tensile strength σ_B

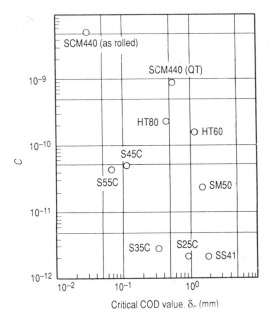

Figure 3.11 Relation between material constant C and critical COD value δ_C

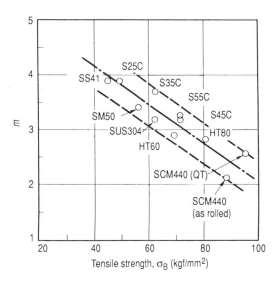

Figure 3.10 Relation between material constant m and tensile strength σ_B

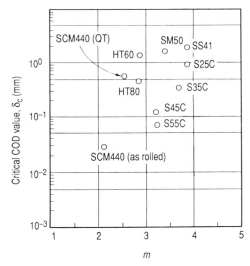

Figure 3.12 Relation between material constant m and critical COD value δ_C

3.4 Final fracture

constants C and m and the critical COD value (Figures 3.11 and 3.12). From the above, it might be thought that there is no correlation between the figures representing toughness and the fatigue crack propagation characteristic.

An abrupt fracture occurs when a force greater than a certain value is applied to a crack or crack-like defect. In most cases, the fracture is a brittle fracture. The fracture surface in this case is a cleavage fracture (see Figures 2.8 and 2.9). The fracture is usually caused by a force acting vertically on the cracked section (mode I). The stress intensity factor in this case is called the critical stress intensity factor and is denoted by K_{Ic}. Since K_{Ic} represents the resistance to fracture, it is also called the fracture toughness value and is considered to be a material constant [5]. Especially in the case where a final fracture is initiated by a fatigue crack, this value is called the fatigue fracture toughness value. In this case, the value is about 10–15% lower than values obtained in the static fracture toughness test. This is because the width of a stretch zone formed at the tip of a fatigue crack [6] is not the same in the two cases.

References

1. Nishida, S. (1976) Doctoral dissertation, 14
2. Schijve, J. (1967) ASTM STP-415, 415
3. Urashima, C., Nishida, S. and Masumoto, H. (1981) *Journal of the Society of Materials Science (JSMS)*, Papers of the Committee on Fatigue, No. 3, 1
4. British Standards Institution (1979) British Standard BS 5767, BSI, London
5. Iron and Steel Institute of Japan (1982) *Data sheet on fracture toughness values of structural steels*
6. Kawasaki, T., Sakamoto, S., Nakanishi, A. and Yokobori, T. (1977) In *Proceedings of the 22nd Symposium on the Strength of Materials and Fracture in Japan*, p. 127, Sendai, Japan

4   Case studies and analysis of failure

4.1 Failures occurring about us

Failures often occur where articles in daily use are so badly damaged, cracked or broken that they are no longer usable. In such cases, however, the causes of failures are usually not investigated but are accepted as the end of service life. This is probably because substitutes are easily available and all articles eventually become unusable because of wear or deterioration. Moreover, we may not have the knowledge and means to investigate the causes of failures. Failure sometimes occurs even where it is important that articles must not break under normal service conditions. Examples of such failures are described below.

4.1.1 Break in the centre shaft of a revolving chair

Figure 4.1 shows a break in the centre shaft of a revolving chair. Figure 4.1(a) is a general view of the revolving chair. Figures 4.1(b) and (c) show the top and bottom of the broken shaft, respectively. Figures 4.1(d) and (e) show enlarged views of (b) and (c), respectively. It can be seen that the break occurred at the stepped portion, i.e. the annular groove which is cut at the outer periphery for the adjustment of chair height. The fracture surface is a typical fatigue fracture surface, and shows that the shaft was under the repeated application of bending stresses. A person sitting on the chair or leaning back in the chair could be seriously hurt if they were to fall from the chair and strike their head because of a break in the centre shaft. It is therefore absolutely necessary to prevent such a failure.

Let us assume the actual service conditions and make simple stress calculations to ascertain whether fatigue failure occurs under the assumed conditions. Assume that the diameter d of the minimum cross section of the centre shaft is 20 mm, the diameter D of the stepped portion is 28 mm, and the radius r of the rounded part of the stepped shaft is 1 mm. Moreover, the material of the shaft is assumed to be S10C which is not heat treated. The S–N curve of this steel is shown in Figure 4.2. The fatigue limit σ_{w0} is 24.0 kgf/mm^2. If the tensile strength σ_B is assumed to be 42 kgf/mm^2, ξ_1 (σ_B), ξ_2 (d), ξ_3 (d/ρ), and ξ_4 [$1-(d/D)$] will be 1.38, 0.86, 0.76 and 0.80, respectively. The notch factor β is then $1 + \xi_1\xi_2\xi_3\xi_4 = 1.72$ [1].

Assuming that a bending moment of 60 kgf × 30 cm is applied as an external force, the stress σ is calculated as 22.9 kgf/mm^2 by substituting $M = 60 \times 300 = 18\,000$ kg mm and $z = \pi/32d^3 = 785.4$ mm^3 into $\sigma = M/z$. When the notch factor β is 1.72, the fatigue limit $\sigma_{w0} = \sigma_w/\beta = 14.0$ kgf/mm^2. Thus, assuming that the original S–N curve can be translated, the life N_f under the repeated application of σ_a of 22.9 kgf/mm^2 is calculated as 7.3×10^4 cycles.

The chair had been in use for about 12 years. If it is assumed that the frequency of stress cycles is 20 cycles per day and the average service ratio is 70%, the total number of cycles, N_T, during this period is $12 \times 20 \times 365 \times 0.7 = 6.13 \times 10^4$ cycles. The order of N_T is equal to that of N_f calculated above.

As a countermeasure to prevent such a break, let us consider the use of a centre shaft made of quenched and tempered S30C steel in which the diameter d of the minimum cross section and the radius r of the rounded part of the stepped shaft are increased to 22 mm and 1.5 mm, respectively, without increasing the diameter D of the stepped portion from 28 mm. The S–N curve of this steel is curve 3 shown in Figure 4.2, and the fatigue limit σ_{w0} is 31.0 kgf/mm^2. If the tensile strength σ_B is assumed to be 65.3 kgf/mm^2, ξ_1 (σ_B), ξ_2 (d), ξ_3 (d/ρ), and ξ_4 [$1-(d/D)$] will be 1.65, 0.88, 0.65, and 0.70, respectively. The notch factor β is then $1 + \xi_1\xi_2\xi_3\xi_4 = 1.66$.

If the external force is assumed to be the same as that described above, the stress σ is 17.2 kgf/mm^2. When the notch factor β is 1.66, the fatigue limit σ_w is 18.7 kgf/mm^2. If the S–N curve is translated in the same way as described above, $\sigma_w = 18.7 > \sigma = 17.2$ kgf/mm^2. Accordingly, it may be said that this shaft will not break. Although a higher safety factor may be necessary to provide for seating for heavier people, the improvement described above will be sufficient from the standpoint of balance between the service life and the production cost of the chair. Namely, a chair which can be used semi-permanently under normal service conditions can be produced by increasing the diameter d and the radius r to 22 mm and 1.5 mm, respectively,

(a)

(b)

(c)

(d)

(e)

Figure 4.1 Break in the centre shaft of a revolving chair (diameter of centre shaft = 20 mm): (a) general view of revolving chair; (b) and (c) top and bottom of broken shaft respectively; (d) enlarged view of (b); (e) enlarged view of (c)

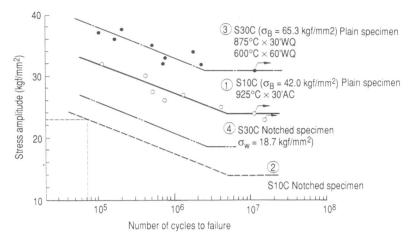

Figure 4.2   *S–N* curves for S10C and S30C

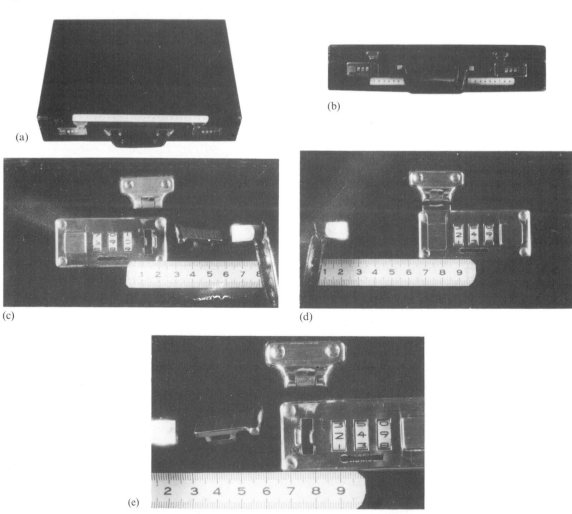

Figure 4.3   Break in the locking metal of an attaché case: (a) general appearance of attaché case; (b) top of case; (c) and (d) enlarged views of (b); (e) enlarged view of (d)

without changing the diameter D from 28 mm and by using quenched and tempered S30C steel instead of S10C.

4.1.2 Break in the locking metal of an attaché case

The broken locking metal of an attaché case is shown in Figure 4.3. Figure 4.3(a) shows the general view of the attaché case which carries papers of A3 size. Figure 4.3(b) shows the top of the case viewed from the locking metal side, while Figures 4.3(c) and (d) are the enlarged views of the locking metal. Figure 4.3(e) is an enlarged view of (d) which clearly shows the broken part of the locking metal. The attaché case is provided with two locking metal components, one on either side. When one of the locking metal components broke, the other was closely checked. As a result, cracking was detected, showing signs of approaching failure. In other words, it could be said that both locking metals components broke at almost the same time. In this attaché case, the spring constant and material of the two locking metal components are the same. When the metal is pushed up to open the case, it is moved upwards and stopped at a certain angle by the action of the spring. The supposition is that the failure of the metal is a fatigue failure caused by continual repetition of the impact force produced when the metal is stopped by the action of the spring at the time of opening. As the repeated stress is assumed to be constant, the number of stress cycles applied to the metal during a given service period can be calculated. Meanwhile, where striation is observed on the fracture surface, both the stress applied to the locking metal and the number of stress cycles can be estimated by measuring the striation distance.

Figure 4.4 shows the results of observation of the fracture surface of the locking metal component by SEM. The thickness of the broken portion is 1.2 mm and cracks are propagating from the upper to the lower side. It is therefore estimated that the fatigue failure was caused by repetition of the impact force produced when the locking metal component was stopped by the action of the spring at the time of opening. Figure 4.4(a) shows the whole fracture surface, while Figure 4.4(b) is an enlarged view of (a). No large artificial defects are observed at the crack initiation point. Figures 4.4(c) and (d) show the surface at a point 0.3 mm from the fatigue crack initiation point, while Figures 4.4(e) and (f) show the surface at a point 0.6 mm from the fatigue crack initiation point. As is apparent from the photographs, the fracture surface can be clearly observed even by an SEM with a relatively low magnifying power, and striation can be clearly observed if an SEM with a higher magnifying power is used. It is obvious from the photographs that the break was caused by the repetition of impact stress in service. As described above, the applied stress and

the number of cycles can be estimated by measuring and analysing the striation distance. For details of the analysis method, see Section 4.2 and ensuing sections.

As a countermeasure to prevent a break in the locking metal, the manufacturer of this attaché case changed the locking metal to a new type. Namely, the manufacturer trebled the thickness at the broken portion, i.e. from 1.2 mm to 3.6 mm, and changed the manufacturing method to solid casting to minimize changes in cross section. As the metal thickness was trebled without changing the width, the modulus of section, z, equal to $1/6 \cdot bt^2$ (where b is the width and t is the thickness), has been increased nine times. Accordingly, as the nominal bending stress is decreased to $1/9$, the possibility of failure has been eliminated. Since the size of the locking metal component is very small, a partial trebling of its thickness will have little effect on the production cost. Complaints from users about failure of the locking metal, subsequent parts replacement, and resultant loss of reputation are more serious for the manufacturer than a slight increase in production cost. In this sense, the countermeasure taken by the manufacturer is reasonable.

References

1. Japan Society of Mechanical Engineers (1982) *Design Handbook of Fatigue Strength of Metals*, Vol. 1, p. 125 (see Appendix 3), Tokyo

4.2 Failure of a crane head sheave hanger

The head sheave hanger of an overhead crane (rating: 5 tf) used in a certain warehouse broke away from its centre. The failure occurred only three months after hanger replacement which was performed during the periodical crane inspection. Since this type of failure may cause an accident resulting in injury or death and many cranes of similar type are in use, the failure was investigated in detail.

4.2.1 Outline of failure

1. Broken part: crane head sheave hanger
2. Manufacturing process: plate SS41→cutting→ heating→bending→final product
3. Duration of service: approximately 3 months
4. Number of cycles to failure: $N_f = 96$ days × 200 cycles/day = 1.92×10^4 cycles

Figure 4.5 schematically shows the crane and the position of failure, while Figure 4.6 shows the dimensions of the crane head sheave hanger. The hanger is

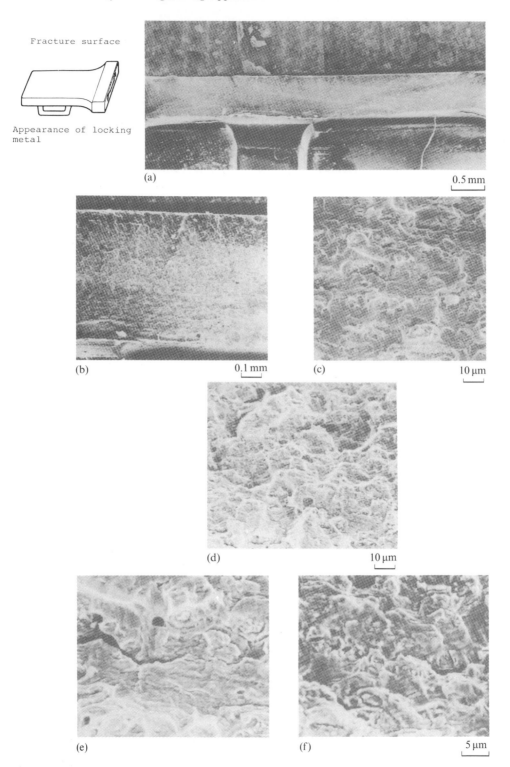

Figure 4.4 Results of observation of fracture surface of locking metal by SEM: (a) whole fracture surface; (b) enlarged view of (a); (c) point 0.3 mm from crack initiation point; (d) point 0.3 mm from crack initiation point; (e) point 0.3 mm from crack initiation point; (f) right-hand side view of (e)

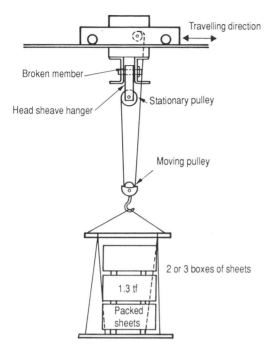

Figure 4.5 Schematic illustration of crane and broken position (rated load: 5 tf)

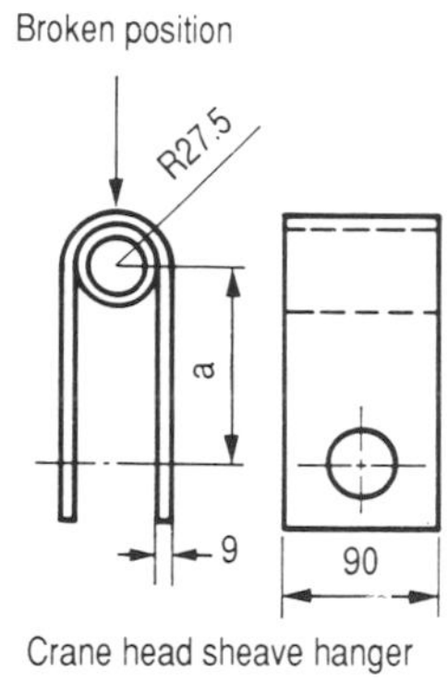

Figure 4.6 Dimensions of crane head sheave hanger

made by heating and bending a plate of thickness 9 mm and width 90 mm. The failure occurred in the centre of the bent portion of the hanger (see Figure 4.6).

4.2.2 *Investigation of cause of failure and analysis*

(a) Microstructure
The sound portion of the hanger has a typical ferrite–pearlite structure with a carbon content of 0.13–0.15%, while the microstructure in the vicinity of the

broken part is coarsened and a Widmanstätten structure is observed [1]. It is therefore highly probable that this part was heated to about 1000–1200°C (photographs are omitted).

(b) Hardness distribution
Figure 4.7 shows the hardness distribution in the vicinity of the fracture initiation point ($H_V(5)$ means that measurement is made under a load of 5 kgf; the same applies hereinafter). The average hardness H_V ($n = 10$ points) in the sound portion is 118. As the hardness at the fracture initiation point is higher by about 50, it is not reasonable to attribute the decrease in fatigue strength at the fracture initiation point to changes in microstructure.

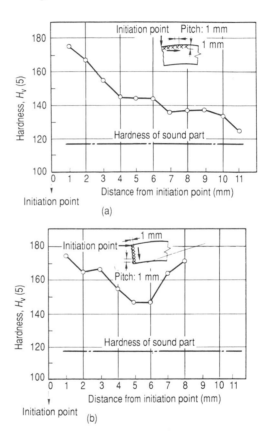

Figure 4.7 Hardness distribution at crack initiation point: (a) in the longitudinal direction; (b) through the thickness

(c) Observation of the fracture surface
Figure 4.8 shows the results of macroscopic and microscopic observation of the fracture surface. It will be noted from Figure 4.8(a), (b) and (c) that this failure is a typical fatigue failure and that cracks are initiated

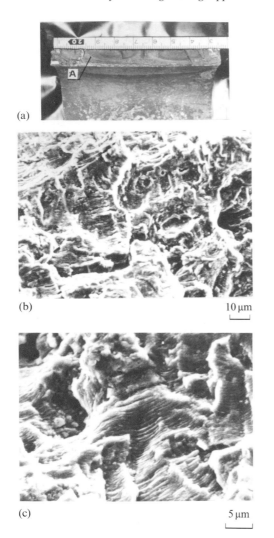

(a)

(b)   10 μm

(c)   5 μm

Figure 4.8 Results of observation of fracture surface of head sheave hanger by SEM: (a) appearance of fracture surface; (b) point A of (a); (c) enlarged view of (b)

and propagated from the outside surface of the bent plate. Striation is also observed and the striation distance at a point 12.5 mm from the fracture initiation point is 2×10^{-4} mm.

(d) Failure analysis

(i) For the ideal loading condition
Figure 4.9 shows the ideal loading condition. Since the effect of frictional force produced between the axle and the head sheave hanger is negligibly small, the stress σ_1 induced in the hanger is uniform within its cross section and can be expressed as:

$$\sigma_1 = \frac{P/2}{bt} = \frac{500/2}{90 \times 9} = 3.1 \text{ kgf/mm}^2 \qquad (4.1)$$

where b is the plate width and t is the plate thickness. Failure will not occur under repeated application of such small stresses, as described later. It is therefore considered that the head sheave hanger was under a loading condition which differs slightly from the ideal condition shown in Figure 4.9.

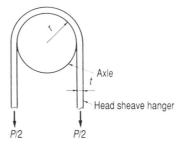

Axle

t

Head sheave hanger

$P/2$ $P/2$

Figure 4.9 Head sheave hanger under ideal loading condition

(ii) Actual loading condition
In the actual loading condition, the contact between the head sheave hanger (1) and the axle (2) is limited over a very narrow range, as shown in Figure 4.10, and it is supposed that this range varies with the magnitude of the applied load. Failure is attributable to the large tensile stress generated at the point A by bending (see Figure 4.10(a)). Namely, it is considered that the actual loading condition was as shown in Figure 4.10(b). To simplify the calculation, let us consider the case where the concentrated load P is applied to the inside at the point A. The maximum bending moment M_{max} is generated at the point where the load P is applied.

$$M_{max} = PR^2 \frac{2-\pi}{a+2\pi R} + \frac{PR}{2} = PR\left[\frac{R(2-\pi)}{a+2\pi R} + \frac{1}{2}\right] \qquad (4.2)$$

Assume that the load P is 5000 kgf, the bending radius R of the head sheave hanger is $r + t/2 = 32.0$ mm, and the length a of the straight portion of the head sheave hanger is 165 mm. Then

$$M_{max} = 5000 \times 32\left[\frac{32(2-\pi)}{165+2\pi \times 32} + \frac{1}{2}\right] = 63\,840 \text{ kgf mm}.$$

The modulus of section, z, is $(1/6)bt^2 = 1215$ mm^3, and the stress σ_A at the point A is

$$\sigma_A = \frac{M_{max}}{z} = 52.5 \text{ kgf/mm}^2$$

Next, let us calculate the stress under the distributed load as shown in Figure 4.10(c). In this case, the load distribution at the contact surface between the head sheave hanger and the axle is not always clear. Accordingly, the load distribution is assumed to be uniform. Since a is 165 mm, which is considerably

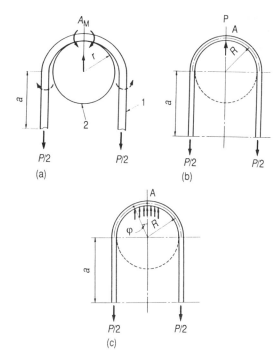

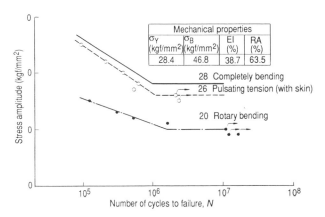

Figure 4.11 *S–N* curves for SS41

σ_a in completely pulsating bending is calculated from the modified Goodman equation:

$$\sigma_a = \sigma_w[1 - (\sigma_m/\sigma_B)] \tag{4.4}$$

where σ_m is the mean stress and σ_B is the tensile strength of the material.

Substituting $\sigma_w = 20.0\,\text{kgf/mm}^2$, $\sigma_m = \sigma_a$ and $\sigma_B = 46.8\,\text{kgf/mm}^2$ into equation (4.4), σ_a is calculated as $14.0\,\text{kgf/mm}^2$, i.e. $\sigma_u = 2\sigma_a = 28.0\,\text{kgf/mm}^2$ in the stress range. This value is reasonable compared with the *S–N* curve of the specimen with mill scale which was obtained by performing the pulsating tension test. In the case of rotational bending, the specimen surface is finished with emery paper ($\phi 7.5\,\text{mm}$), and therefore the effect of the surface condition should be taken into consideration in the case of pulsating bending. However, the *S–N* curve (pulsating bending) obtained by converting the results of the rotational bending fatigue test according to the modified Goodman equation is reasonable compared with that for pulsating tension (with mill scale). Accordingly, a σ_u value of 28.0 kgf/mm² was adopted here and the *S–N* curve was drawn parallel to that for the case of pulsating tension. Fatigue failure occurs when the repeated stress is as shown below:

$$\sigma_A' = 40.0\ \text{kgf/mm}^2 > \text{fatigue limit of material } \sigma_u$$
$$= 28.0\ \text{kgf/mm}^2$$

Moreover, when the actual load $P_a = 3.9\,\text{tf}$ (three boxes, $1.3\,\text{tf} \times 3$) instead of 5 tf, the repeated stress σ_A'' is 31.2 kgf/mm² and fatigue failure will occur. When the repeated stress σ_A' is 40.0 kgf/mm² and σ_A'' is 31.2 kgf/mm², the number of cycles to failure will become 3×10^4 and 4×10^5 cycles, respectively. The order of these values is the same as that of the actual number of cycles $N_f = 1.92 \times 10^4$ cycles. In the case of cranes, it is well known that stresses larger than expected are induced at the time of lifting, travelling and stopping. If these stresses are taken as the dynamic loads, the dynamic load factor will usually be 1.2–1.5.

Figure 4.10 (a) Actual loading condition. (b) Concentrated loading ($R > r$). (c) Distributed loading ($R > r$)

greater than the other sizes, the bending moment M' is calculated as shown below.

$$M' = r \cdot r\varphi\omega - \int_0^\varphi r\,\sin\,\theta\,\,r\omega\mathrm{d}\theta = r^2\varphi\omega\,\cos\,\varphi$$

From the locus of contact between the head sheave hanger and the axle, assume that $\psi = \pi/4$ and $\omega = P/(2r\psi)$. As r is 27.5 mm and P is 5000 kg, M_{max} is calculated as:

$$M_{max} = \frac{Pr}{2} \cdot \cos\,\varphi = 48\,613\,\text{kgf mm}$$

where $\psi = \pi/4$. The maximum tensile stress σ_A' at the point A under the distributed load is

$$\sigma_A' = \frac{M_{max}'}{z} = \frac{48\,613}{1215} = 40.0\,\text{kg/mm}^2 \tag{4.3}$$

Even in the case of distributed load as shown in Figure 4.10(c), the hanger is subjected to repeated application of stress of 40.0 kgf/mm² in the stress range. As this value is considerably greater than the yield stress of the material, the head sheave hanger is plastically deformed during actual service. It is therefore estimated that the actual repeated stress is lower than 40.0 kgf/mm².

The results of the fatigue test of SS41 are shown in Figure 4.11 in terms of the *S–N* curve [2]. From the results of the rotational bending fatigue test, the stress

4.2.3 *Fracture mechanics approach*

When the bending moment is applied as shown in Figure 4.10(a), the number of cycles to failure and the repeated stress are calculated by making an approximation as shown in Figure 4.10(b). For simplification of calculation and for practice, the stress and the number of cycles in the case of outward bending of the surface were calculated by converting the results of the tensile fatigue test of the specimen with a surface crack. If an exact calculation is required, it is desirable to calculate the values directly by using the stress intensity factor obtained in the case of outward bending of the surface with a crack, although the calculation becomes somewhat complicated. An example is described in Section 4.11.

$$\frac{\mathrm{d}l}{\mathrm{d}N} = C(\Delta K)^m \tag{4.5}$$

The stress intensity factor in the case where a tensile stress is applied to the plate with a surface crack is given by the following equation (see Figure 4.12)

$$\Delta K = \sigma_0 \sqrt{\left/\left(2B \tan \frac{\pi l}{2B}\right)\right.} \tag{4.6}$$

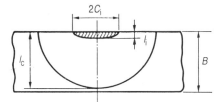

Figure 4.12   Assumptions for calculation

Substituting equation (4.5) into equation (4.6) and assuming that m is nearly equal to 4, $\mathrm{d}N$ is integrated. Then

$$N_c = \int_{l_i}^{l_c} \frac{\mathrm{d}l}{C(\Delta K)^4} = \frac{1}{2\pi C \sigma_0^4 B}$$
$$\times \left(\cot \frac{\pi l_i}{2B} - \cot \frac{\pi l_c}{2B} - \frac{\pi}{2B}\right)(l_c - l_i) \tag{4.7}$$

As B is assumed to be larger than l_i,

$$N_c \fallingdotseq \frac{1}{\pi^2 C \sigma_0^4}\left(\frac{1}{l_i} - \frac{\pi^2}{4B}\right) \tag{4.8}$$

Since the plate surface is bent outwards in this case, it may be considered that $\Delta K \rightarrow (1/2)\Delta K$, i.e. $\sigma_0 \rightarrow (1/2)\sigma_0 = 20$ kgf/mm^2 [3]. When C is assumed to be 7.54×10^{-12}, l_i, 0.5 mm (see Section 3.3), and B, 8.5 mm (B is 9 mm in practice but it is apposite in this case to assume that l_c to failure $= 8.5$ mm $= B$):

$$N_c = 1.44 \times 10^5 \text{ cycles} \tag{4.9}$$

Although the initiation crack depth l_i is assumed to be 0.5 mm, the number of cycles to failure will be about twice the value calculated above because the initial crack does not exist in practice.

$$N_f = 2.9 \times 10^5 \text{ cycles} \tag{4.10}$$

Next, the nominal stress acting on the head sheave hanger is calculated. From Figure 4.6, the striation distance is approximately $2 \, \mu$m (point A in Figure 4.6(a)). The distance is taken as about $1 \, \mu$m by converting it into the distance at the deepest point on the same semi-ellipsoidal contour line. The stress intensity factor K_I in this case is nearly equal to 100 kgf/mm$^{3/2}$ [4]. Substituting $l = 6$ mm and $B = 9$ mm into equation (4.6):

$$\sigma_0 = 17.9 \text{ kgf/mm}^2 \tag{4.11}$$

That is, σ_{u0} of about 35.8 kgf/mm^2 is applied in the stress range. This agrees with the calculation results described above.

4.2.4 *Summary and countermeasures*

On the basis of the results described above, it may be concluded that the failure of the head sheave hanger is attributable to the fatigue caused by the stresses developed under the actual loading conditions which were not taken into consideration at the design stage. As the failure of such a member may cause an accident resulting in injury or death, the prevention of such failures is absolutely imperative. All designers should pay special attention to this point.

As a countermeasure, the plate thickness is increased. According to calculations, the plate thickness should be greater than 12 mm. Whether this thickness is sufficient or not is open to question because the thickness is dependent on the condition of contact between the head sheave hanger and the axle. As a precautionary measure, it is most desirable to design the head sheave hanger so that bending is not brought about (see Figure 4.13).

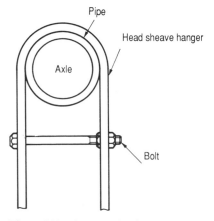

Figure 4.13   An example of a countermeasure

References

1. Nisizawa, Y. and Sakuma, T. (1979) *Photographs of Metal Structure – Steels*, 3 or 56, Maruzen, Tokyo
2. Nippon Steel Corporation (1970) Technical data
3. Okamura, H. (1976) *Introduction to Linear Fracture Mechanics*, p. 165, Baifukan, Tokyo
4. Kitagawa, H. and Koterazawa, R. (1977) *Fractography*, Baifukan, Tokyo

4.3 Failure of wire rope [1,2]

4.3.1 Outline of failure

Wire rope is used for the haulage of men and materials and as structural members in a wide variety of industries, such as machinery, construction, shipping, fisheries, forestry, mining, cables and elevators. Wire rope is one of the representative steel products from the standpoints of versatility, importance and safety.

Accidents caused in the past during crane operations are classified in Table 4.1. Although the percentage of accidents attributable to the failure of wire rope is not clear, it can easily be understood that not only workers engaged in haulage but also other general workers will be involved in an accident should the wire rope break.

Table 4.1 *Accidents in crane operations**

Classification	Number of cases	Percentage
Operator	14	9
Porter	104	66
General workers	39	25

* From an investigation by the Labour Ministry.

In a large proportion of instances the failure of wire rope is attributable to the combined action of various factors, such as wear, wire breakage, corrosion, deformation, and eccentricity. Of these causes, wire breakage is the most dangerous, and is said to be mostly attributable to fatigue [3]. However, little work has been done on this point [4]. In particular, very little work has been directed towards the fractographic investigation of wire rope failures.

As is well known, wire rope is composed of many element wires laid together. In this sense, wire rope is a kind of structure. During service, wire rope comes under various repetitive stresses, such as tensile stress in the axial direction, bending and compressive (secondary bending) stresses that develop when it passes through the sheave, and localized float bending (tertiary bending) stress. Because of the combined action of repeated stresses, wear, fretting fatigue [5], etc. the strength of wire rope deteriorates gradually from the very beginning of service. It is considered that wire rope does not have the fatigue limit which is observed in general steel products. Namely, the remaining service life of wire rope decreases as the duration of service increases. Moreover, unexpectedly short life or breakage under low load may occur occasionally.

With the aim of preventing abnormal failure, extending service life and promoting the maintenance of safety by investigating the broken wire rope in detail, four instances of wire rope failure are reported below [1,2].

4.3.2 Analysis of failure

(a) Case 1: break from an equalizer sheave

(i) Outline of break

When a 30-tf crane lifting a product (15.6 tf) travelled about 3 m to load a truck, the main wire rope broke and the product fell from a height of 2 m.

1. Crane capacity: 30 tf
2. Wire rope: $\phi24$ mm, $6 \times F_i(29)$, length 124 m
3. Lifting load at the time of the break: product 15.6 tf + C-hook 1.3 tf, total 16.9 tf
4. Wire rope replacement: replaced three times – after use for periods of 11 months, 13 months and 9 months (the particular wire rope that broke on this occasion)
5. Wire rope inspection: visual inspection for breakage once a month
6. Breakage point: the equalizer sheave which is in contact with the stationary pulleys used for adjustment of winding between the left and right drums
7. Other details: the broken wire rope cannot be inspected unless the wire rope is completely removed.

The crane and the broken point of the wire rope are shown schematically in Figure 4.14.

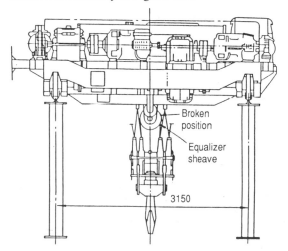

Figure 4.14 Schematic representation of crane and broken wire rope

(ii) Results of observation and test

1. Structure of wire rope: $6 \times F_i(29)$, $(6 \times (1 + 7 + 7 + 14))$, with hemp core, diameter of outer element wire: 13.9 mm, diameter of middle element wire: 1.60 mm, nominal diameter of wire rope: 24 mm, bare, red grease, ordinary right-hand lay, type **B** (180 kgf/mm² class), standard breaking load: 34.7 tf.

2. Tensile test of element wire: that part of the element wire which was broken in the tensile test is shown in Figure 4.15. A reduction in area is only barely observable in the outer element wire, as shown in Figure 4.15(a), while a considerable reduction in area is observed in Figure 4.15(b), (c) and (d).

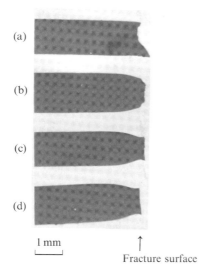

Figure 4.15   Broken part of element wire in tensile test: (a) outside element wire of broken wire rope; (b) middle element wire of broken wire rope; (c) outside element wire of new wire rope of the same type; (d) middle element wire of new wire rope of the same type

3. Observation of fracture surface: a considerable reduction in area which may be attributable to fatigue failure is observed in the fracture surface of the outer element wire of the broken wire rope. Figure 4.16 shows an example of the results of observation of the fracture surface by SEM. Several cracks are seen on the worn outer surface. The cracks propagated and were connected with each other. The final fracture surface is a ductile fracture surface and a dimple pattern is observed (Figures 4.17 and 4.18). However, striation which is characteristic of fatigue failure is not observed on this fracture surface.

(iii) Estimation of the cause of failure

Apparently, the wire rope does not move at all in the vicinity of the equalizer sheave. In practice, however,

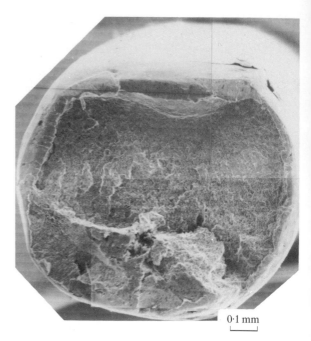

0·1 mm

Figure 4.16   Example of the fracture surface of outside element wire observed by SEM

the element wires of the wire rope are considerably abraded at this point. It is therefore supposed that the wire rope moves (including minor slips) in the longitudinal direction or rotates under load in the vicinity of the sheave.

1. Repeated stress: the lifting load is 15–17 tf, including the weight of the C-hook. For this crane, the dynamic load factor is assumed to be 1.2–1.3 because of excessive vibration, and the load is assumed to be about 20 tf. The cross-sectional area A of the wire rope is 247 mm² and the tensile stress $\sigma_T = P/A = 10.1$ kgf/mm². The stress σ_b created by bending is calculated as shown below:

$$\sigma_b = E_r\, \delta/D = 41.7 \text{ kgf/mm}^2 \tag{4.12}$$

in which δ is the diameter of the outer element wire, equal to 1.39 mm, D is the diameter of the equalizer sheave, equal to 300 mm, and E_r is the modulus of elasticity of the wire rope, equal to 9000 kgf/mm².

2. Estimation of the number of cycles to failure: the number of cycles to failure can be estimated from the stresses calculated in item 1 above. In many cases, however, the number of cycles to failure is calculated in relation to D/δ or D/d in which D is the wheel diameter, δ the diameter of the outer element wire, and d the nominal diameter of the wire rope. Since D is 300 mm and d is 24 mm, D/d is 12.5.

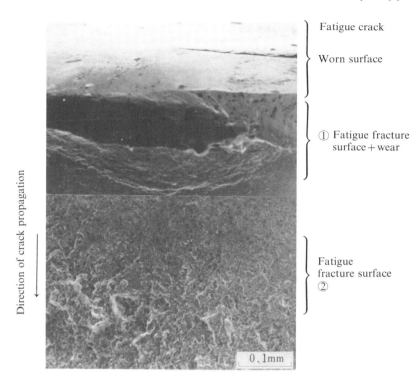

Figure 4.17   Worn surface, fatigue crack and fatigue fracture surface (enlarged view of Figure 4.16)

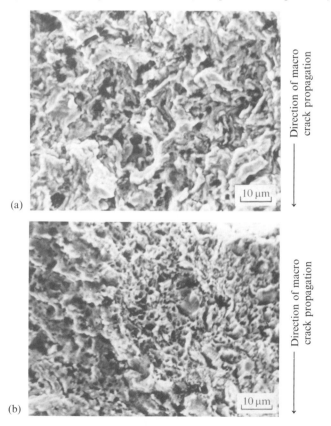

Figure 4.18   Example of fracture surface observed by SEM: (a) fatigue fracture surface; (b) final ductile fracture surface (dimple pattern)

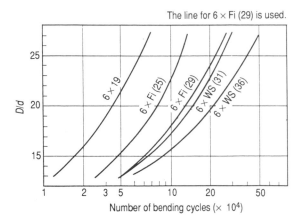

Figure 4.19 Number of bending cycles to 10% break of various ropes (180 kgf/mm² class, ordinary lay, safety factor 6)

Figure 4.19 shows the relation between D/d and the number of bending cycles to 10% break for various wire ropes. The number of cycles to 10% break in the case of D/d equal to 12.5 is 4×10^4.

3. Lifting frequency of product: the number of stress cycles, N_f, actually applied to the wire rope is calculated as shown below:

$$N_f = 10 \text{ times/hour} \times 24 \text{ hours/day} \times 30 \text{ days/month} \times 9 \text{ months} \times 0.7 \text{ (operation rate)}$$
$$= 4.5 \times 10^4 \text{ cycles}$$

The number of cycles thus calculated exceeds the number of '4×10^4 cycles' which was calculated in item 2 above.

(iv) Conclusion and countermeasures

No abnormalities are observed in the properties of the material of the element wires. It may therefore be concluded that fatigue failure occurred in the outer element wire because of the wear and fatigue caused by repeated application of tensile and bending stresses to the wire rope at the equalizer sheave, and that this failure spread to other wires. Two countermeasures can be taken to prevent this type of failure.

1. The service life of the wire rope is prolonged by increasing the diameter of the equalizer sheave.
2. The entire wire rope should be closely inspected. That is, the point of contact between the wire rope and the equalizer sheave should be closely inspected by displacing the winding amount between the left and right drums, although this method may take time.

(b) Case 2: break from the portion passing through the sheave

(i) Outline of break

The wire rope of a 0.5-tf crane broke suddenly when the crane travelled a distance of about 0.5 m to haul a product to on adjacent pile. The break is shown schematically in Figure 4.20. The record of wire rope replacement is shown in Table 4.2. The tonnage handled by this wire rope is smaller by about 30 000 tf (24% on average) than past records show. The future rope maintenance system is dependent upon whether the break is an early abnormal break or is attributable to variations in the fatigue life of the wire rope.

1. Crane capacity: 5 tf, winding drum diameter: 346 mm, sheave diameter: 240 mm
2. Wire rope diameter: 12 mm, 6×24, steel grade SWRA62-A, breaking load: 7.6 tf (standard load 7.4 tf)

Table 4.2 *Records of wire rope replacement**

No.	Period (month)	Handled tonnage (tf)	Number of cycles, N	Remarks
1	9	139 800	46 600	
2	8	142 000	47 333	
3	10	131 000	43 666	
4	4	41 700	–	Wound into drum
5	8	105 000	35 000	Present break

*The total number of cycles was calculated on the assumption that the lifting load is 3 tf/time on the average.

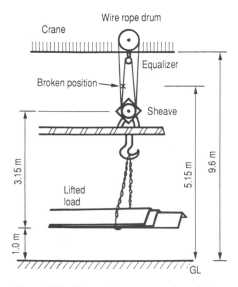

Figure 4.20 Schematic representation of wire rope failure

3. Lifting load at the time of break: 3.0 tf
4. Wire rope replacement: shown in Table 4.2
5. Frequency and method of inspection of wire rope: the same as those in case 1
6. Broken part: the part which passes through the sheave, as shown in Figure 4.21

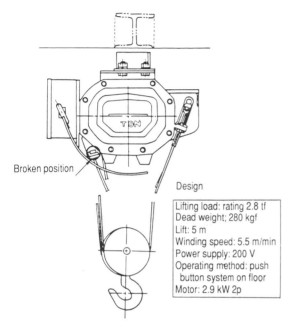

Broken position

Design

Lifting load: rating 2.8 tf
Dead weight; 280 kgf
Lift: 5 m
Winding speed: 5.5 m/min
Power supply: 200 V
Operating method: push button system on floor
Motor: 2.9 kW 2p

Figure 4.21 Equipment outline and position of failure in wire rope

7. Other details: the broken wire rope received a monthly inspection five days before the occurrence of the break. This visual inspection did not reveal the breakage of element wires and the wire rope diameter was 11.9 mm (nominal diameter of new wire rope: 12.0 mm). No specific abnormalities were reported.

(ii) Results of observation and test
1. Appearance of wire rope: the broken wire rope is shown in Figure 4.22. The breakage of many element wires is observed not only in the broken part of the wire rope but also in the unbroken part. However, wire breakage is not seen in the wire rope in the vicinity of the winding drum.
2. Tensile test of elements wires: the element wires (nominal diameter: 0.67 mm) taken from the wire ropes were subjected to tensile test after straightening and rolling of the portion to be gripped. The results of test are shown in Table 4.3. The breaking stress of the elements wire (average nominal stress) was considerably decreased to 128.5 kgf/mm^2 in the vicinity of the broken part. However, the breaking stress of the wire in the vicinity of the drum was 180.0 kgf/mm^2 and that of new wire was 177.5 kgf/mm^2. At the unbroken part, the breaking stress was 142.2 kgf/mm^2 which was intermediate between the above two values. This may be attributable to the reduction in the wire diameter because of wear and the localized initiation of fatigue cracks.
3. Wear and wire breakage: the worn and broken condition of the wire rope are shown in Figure 4.23. The wear losses of the wire rope and the element wires are shown in Table 4.4. The wear of the wire rope is one-sided and occurred mostly at the outer periphery in contact with the sheave. The ratio of wear loss of broken wire to the diameter of new wire rope (actually measured diameter $d_0 = 12.80$ mm) is 12.0%. This wear loss is quite high. However, the ratio of wear loss to the nominal diameter (12.0 mm) is only 4.2%, satisfying Article 215 of the Crane Safety Regulations (which stipulates that the wire rope should be replaced when its wear loss exceeds 7%). When the wire rope was bent slightly by an amount corresponding to the sheave radius, the number of wires broken in the vicinity of the broken part increased sharply.
4. Observation of fracture surface: fatigue cracks initiated at the worn portion and propagated, resulting in a break (photographs are omitted because they are similar to Figures 4.16 and 4.17 in case 1.)

(iii) Estimation of cause of break
1. Repeated stress: lifting load $P = 3000$ kgf, dynamic load factor $K_D = 1.5$, number of wire ropes, $n = 8$, cross-sectional area of wire rope, $A = 40.72$ mm^2, tensile stress developed by lifting load, $\sigma_T = 13.8$

← New wire rope

← Unbroken part

← Broken part

Broken wire rope

Figure 4.22 Appearance of broken wire rope

Table 4.3   *Results of tensile test of element wire (nominal diameter, 0.68 mm)*

Type	Symbol	Breaking load (kgf)	Nominal stress (kgf/mm²)	Mean value $\bar{x}$ (kgf/mm²)	Ratio to new wire rope
Broken wire rope					
In the vicinity of broken point	No. 1	42.2	119.7		
	2	46.0	130.5	128.5	0.72
	3	47.7	135.3		
Unbroken part	No. 1	51.4	145.8		
	2	50.0	141.8	142.2	0.80
	3	49.0	139.0		
In the vicinity of drum	No. 1	61.0	173.0		
	2	64.9	184.1	180.0	1.01
	3	64.5	183.0		
New wire rope	No. 1	65.0	184.4		
	2	61.3	173.9	177.5	1.0
	3	61.4	174.2		
Specification		55.4–65.0	157.1–184.4	–	–

Broken wire rope (approximately 1 m from broken section)

Unbroken wire rope

Wire rope in the vicinity of drum

New wire rope

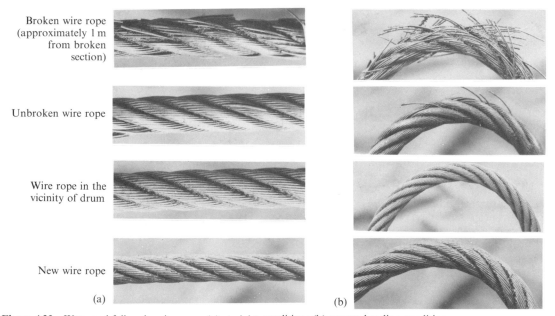

(a)   (b)

Figure 4.23   Wear and failure in wire rope: (a) straight condition; (b) reverse bending condition

Table 4.4.   Wear and break in wire ropes

Type	Wear of wire rope* (mm) d_1	d_2	Wear loss (%) $\left(1 - \dfrac{d_1}{d_0}\right) \times 100$	Wear of outside element wire	Maximum number of broken element wires per pitch
Broken wire rope in the vicinity of broken point	11.50	12.10	10.2	▬▬▬▬	45 wires
Unbroken part	11.50	12.00	10.2	▬▬▬	10 wires
In the vicinity of drum	12.10	12.40	6.3	▬▬▬	0
New wire rope	12.80	12.80	–		0

$d_0 = 12.8$ mm, measured by vernies caliper.

kgf/mm², diameter of element wire, $\delta = 0.6$ mm, sheave diameter $D = 240$ mm, modulus of elasticity of wire rope, $E_r = 9000$ kgf/mm², bending stress developed at the time of passage through the sheave, $\sigma_b = 22.5$ kgf/mm²

2. Estimation of the number of cycles to break: although the tensile strength of each element wire varies slightly, the approximate number of cycles to 10% break can be obtained from the '$6 \times F_i(25)$' line shown in Figure 4.16 in the same way as in case 1. Namely, the number is 7.5×10^4 at $D/d = 20$. However, as the wire rope was bent twice each time the load was lifted, the number of bending cycles to break, N_f, is calculated as shown below on the basis of values shown in Table 4.2:

$$N_f = 3.5 \times 10^4 \times 2 = 7.0 \times 10^4$$

Namely, both values are nearly equal.

(iv) Conclusion and countermeasures

On the basis of the results described above, it may be said that the break in the wire rope is not an abnormal break but is due to wear and fatigue caused by repeated loads. The countermeasures for such a break are described below.

1. The sheave diameter is increased. If the diameter D of the sheave now in use is increased from 240 mm to 300 mm, D/d becomes 25, and therefore the current life is increased 1.6 times.
2. The wire rope maintenance system is improved. The wire rope is replaced when its wear loss exceeds 7% of the diameter of new wire (the actually measured value). Namely, acceptance inspection of the wire rope is necessary. For breakage of element wires, the wire rope is replaced if more than six element wires are broken per pitch when the wire rope is slightly bent by an amount corresponding to the radius of the sheave.

Case 3: break of wire rope from the end stopper

(i) Outline of break

A wire rope used for a 2.8-tf hoist crane broke from the end stopper when a product of weight 2 t was being loaded into a truck by the crane. The crane and the broken point of the wire rope are shown in Figure 4.21.

1. Crane capacity: 2.8 tf
2. Wire rope diameter: 10 mm, 6×24
3. Lifting load at the time of the break: 3.4 tf
4. Duration of use of wire rope: 505 days (17 months), number of cycles to break, $N_f = 505$ days $\times 20$ cycles/day = 10 100 cycles; this number is far less than that for ordinary wire rope
5. Frequency and method of wire rope inspection: the same as for case 1
6. Break point: at the end stopper of the wire rope; the wire rope broken at the end stopper is shown in Figure 4.24
7. Other details: the wire rope of a crane usually breaks at the portion which passes through the sheave. A break at the end stopper is a very rare occurrence. The first suspected cause of such a break is an imperfect casting condition. As is apparent from Figure 4.24(b), showing the broken wires in the longitudinal cross section of the socket, the wires usually come out of the socket if the casting condition is imperfect. In case 3, the majority of the wire elements break from the portion bound with wire, and therefore the break in case 3 cannot be attributed to an imperfect casting condition.

(ii) Results of observation and test

1. Structure of wire rope: JIS No. 4, type A, ordinary right-hand lay, nominal diameter: 10 mm, 6×24, diameter of element wire: 0.56 mm (tensile strength: 180 kgf/mm² class), standard breaking load: 5.02 tf

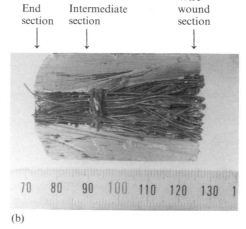

End section   Intermediate section   Wire-wound section

(a)   (b)

Figure 4.24   Breakage of wire rope at end stopper: (a) breakage; (b) vertical cross section of socket

Table 4.5(a)   *Properties of broken wire rope*

Chemical composition (Wt%)				
C	Si	Mn	P	S
0.62	0.25	0.47	0.012	0.013

Mechanical properties		
Sampling position	Tensile strength σ_B (kgf/mm^2)	Hardness, H_v(0.5)
Vicinity of broken point	184.7	–
Unbroken part	185.5	538.6

Fibrous ferrite and pearlite
Microstructure   No abnormality

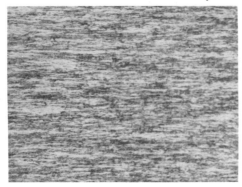

50μm

Table 4.5(b)   *Mechanical properties of wire rope*

Sampling position	Strand No. O	Break load (kgf)	Tensile strength* (kgf/mm^2)	Hardness number, H_v (0.1)
Sound portion	1	275.0	136.8	
		284.0	141.3	Mean value of hardness at 7 points $\bar{x} = 416.6$
	2	277.0	137.8	
		302.0	150.2	
	3	287.0	142.7	
		303.0	150.7	
Vicinity of broken portion	1	268.0	133.3	
		202.0	100.5	
	2	233.0	115.9	
		170.0	84.6	
	3	235.0	116.9	Same as above, $\bar{x} = 421.0$
		277.0	137.8	
	4	265.0	131.8	
		172.0	85.5	
	5	256.0	127.3	
		193.0	96.0	
	6	248.0	123.3	
		178.0	88.5	
Specification	JIS No. 4 Type A	271.4– 321.7	135.0–160.0	–

*A nominal diameter of 1.6 mm was used.

2. Investigation of properties: the results of an investigation of the properties of wire rope are shown in Table 4.5. Both the chemical composition and the mechanical properties are normal. The microstructure consists of fibrous ferrite and pearlite.
3. Analysis of fracture surface: the majority of the fracture surfaces of the wires broken from the portion bound with wire (the number of wires broken in this portion is 60, accounting for 43% of the total number of broken wires) are comparatively smooth. However, a reduction in area is observed in the fracture surfaces of many wires broken at the socket end (about 30 wires, accounting for 21% of the total number of broken wires), while both smooth surfaces and a reduction in area are observed in the fracture surfaces of the wires broken in the intermediate portion (see Figure 4.25). Figure 4.26 shows an example of the fracture surface of a wire element which was broken at the portion bound with wire. A beach mark (shell pattern) which is peculiar to fatigue is observed, and therefore this fracture surface is a typical fatigue fracture surface. The final fracture surface is a ductile fracture surface and dimples are observed. In the wire elements which were broken at the socket end, dimples are observed all over the surface and a considerable reduction in area is seen visually. This example is shown in Figure 4.27.

(iii) Estimation of cause of break
No abnormalities are observed in the properties of the material of the wire rope but the life of the wire rope is much shorter than that of conventional wire rope.

Broken position

Portion bound with wire   Intermediate portion   End portion

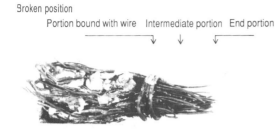

Broken position	Number of broken wires	Features of fracture surface
Portion bound with wire	About 60 wires	Comparatively flat fracture surface
Intermediate portion	About 50 wires	Both flat fracture surface and reduction of area
End portion	About 30 wires	Reduction of area in many fracture surfaces

Figure 4.25 Breakage of element wire in the socket and features of fracture surfaces

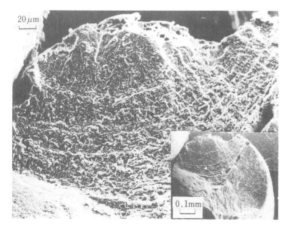

Figure 4.26 Example of fracture surface of element wire broken at the portion bound with wire

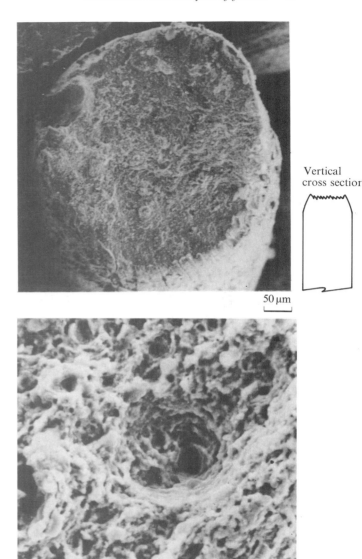

Vertical cross section

Figure 4.27 Fracture surface of element wire broken at the socket end

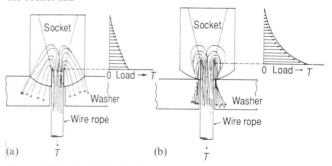

Figure 4.28 Load distributions to element wires in the socket using (a) spherical and (b) flat washers

Moreover, it was found that the wire rope was used with the socket washer placed with the wrong side up. On the basis of these facts, the cause of the break in the wire rope is estimated as described below.

Figure 4.28 qualitatively shows the effect of the socket and washer on the load distribution to the element wires. When a conventional spherical washer is used (Figure 4.28(a)), the load applied to the element wires in the socket is distributed quite uniformly.

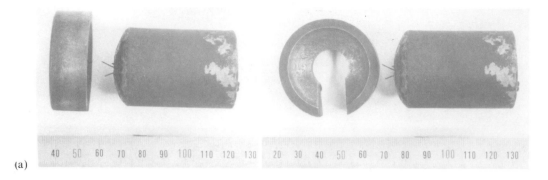

(a)

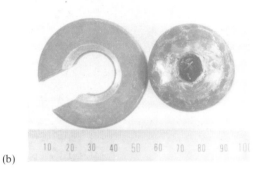

(b)

Figure 4.29 Installation of socket washer: (a) correct; (b) incorrect

When a flat washer is used (as in this case) (Figure 4.28(b)), however, the load is concentrated in the contact area between the washer and the socket, and therefore a large load is applied to the portion bound with wire. Moreover, compared with the case in which a spherical washer is used, an uneven load is apt to be applied to the wire rope when the plain washer is used, increasing the possibility of breakage at the end stopper. Figure 4.29 shows the condition of the installed socket washer. In Figure 4.29(b), ring-shaped marks made by the contact between the socket and the washer are seen around the centre hole on the bottom side of the washer. Namely, it is apparent that the washer was placed with the wrong side up.

(iv) Conclusion and countermeasures
Summarizing the above description, it may be concluded that the main cause of wire rope failure is the concentration of load at the portion bound with wire in the socket due to the incorrect use of the washer. It is considered that fatigue failure occurred at this point and spread, causing the breakage of element wires one after another and finally resulting in the failure of the wire rope. It is also estimated that frequent use of the wire rope under the overloaded condition considerably shortened the life of the wire rope [6,7]. The countermeasures are described below.

1. The socket washer is used in the correct way. If possible, a lubricant is applied to the contact area between the socket and the washer.
2. The shape of the socket is improved (for example, the contact area is increased) to ensure uniform load distribution to the element wires in the socket.
3. A load limiter is installed to prevent overloading.

(d) Case 4: break in nearly new wire rope

(i) Outline of break
During loading of a steel structure (2330 tf) into a barge (at a loading speed of 0.4 m/min), the wire rope of a 15-tf winch broke when the structure was pulled on board through a distance of about 34 m. At the time, the broken wire rope writhed on the barge deck like an injured snake. This condition is shown schematically in Figure 4.30.

1. Winch capacity: rating 15 tf (20 tf maximum)
2. Wire rope: ϕ28 mm, 6 × 24, breaking load: 39.8 tf
3. Load at the time of break: tension 20 tf
4. Wire rope replacement: replaced after use for four years and five months. The broken wire rope is the rope purchased as a spare at that time. The rope broke only five hours after the start of use. The number of cycles to failure, N_f, was $5 \times 20 = 100$ cycles maximum.

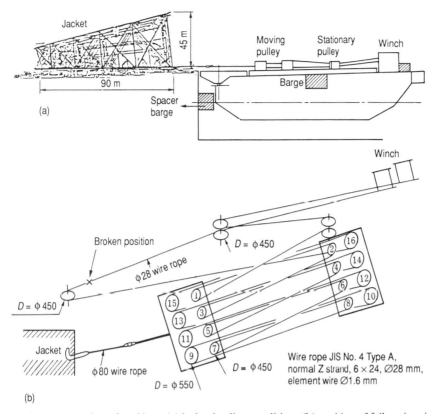

Figure 4.30 Schematic representation of accident: (a) jacket loading condition; (b) position of failure in wire rope

5. Frequency and method of wire rope inspection: the wire rope was visually inspected for breakage of element wires before use.
6. Broken portion: the portion which passes through the sheave (see Figure 4.30).
7. Other details: Although the wire rope was purchased four years and five months ago, it was stored on board ship in a well-greased and neatly wrapped condition. Accordingly, the rope was nearly new. Needless to say, the outside diameter (28 mm) of the wire rope remained unchanged.

(ii) Results of observation and test

1. Wire rope: JIS No. 4, type A, ordinary right-hand lay, nominal diameter: 28 mm, 6 × 24, diameter of element wire: 1.58 mm, standard breaking load: 39.3 tf
2. Appearance of broken wire rope: the broken wire rope is shown in Figure 4.31. As the wire rope was nearly new, it was free from wear. However, close observation revealed red rust here and there. More detailed observation revealed considerable partial reduction in the diameter of the element wires.
3. Investigation of the properties of the material of the wire rope: the microstructure of the longitudinal

cross section of the element wire is shown in Figure 4.32. No abnormalities were detected in the sound or broken portions. The mechanical properties of the element wires are shown in Table 4.5. The tensile strength of the respective wires is 143.3 kgf/mm^2 on average in the sound portion but varied in the range 135.0–160.0 kgf/mm^2 in the vicinity of the broken portion, the average being 111.8 kg/mm^2, which deviates considerably from the standard value of 135.0–160.0 kgf/mm^2. The hardness in the vicinity of the broken portion is nearly equal to that in the sound portion. However, the tensile strength of almost all the element wires is lower than the standard value because of the reduction in diameter resulting from corrosion. Figure 4.33 shows the longitudinal cross section of the element wire. A reduction in diameter to less than half of the initial diameter is observed locally. This tendency is particularly marked in the vicinity of the broken portion.

(iii) Estimation of cause of break

The fracture surfaces of the element wire of the broken wire rope are roughly divided into two types, as shown in Figure 4.34. One is the cup and cone type tensile

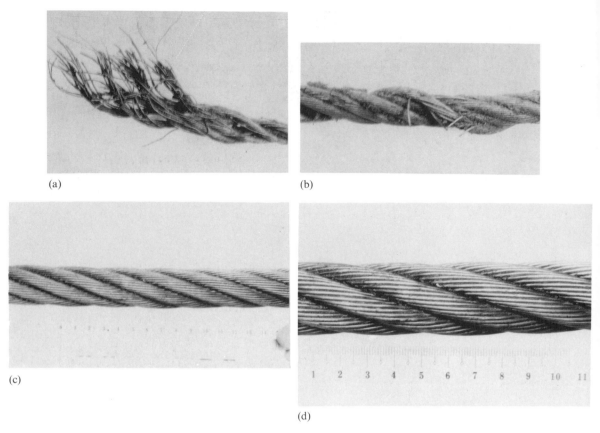

(a)

(b)

(c)

(d)

Figure 4.31   Appearance of broken wire rope: (a) broken portion; (b) vicinity of broken portion; (c) sound portion (degreased with acetone); (d) enlarged view of (c)

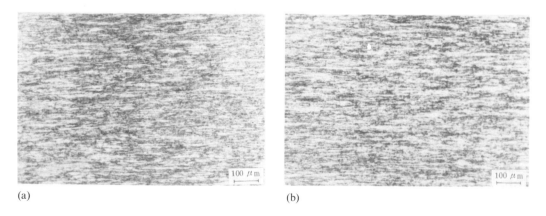

(a)

(b)

Figure 4.32   Microstructure of vertical section of element wire: (a) sound portion; (b) vicinity of broken portion

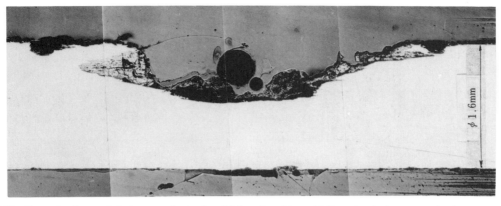

Figure 4.33 Example of corroded element wire in the vicinity of position of failure (vertical cross section)

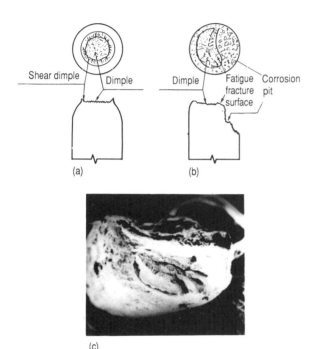

Figure 4.34 Types of fracture of broken wire rope:
(a) A-type fracture; (b) B-type fracture (occurs frequently in outside element wire); (c) example of B-type fracture

ductile fracture (type A) and the other is of the type (type B) in which several fatigue cracks propagate from the portion where the thickness has decreased due to corrosion (the fatigue fracture surface may not be observed in some cases), resulting finally in ductile fracture. It is considered that the failure of a wire rope begins with a type B fracture, leading to a type A fracture finally. In other words, it is considered that plastic fatigue plus ductile fracture, partly direct ductile fracture, are caused by the application of tensile

stress (created by tension) and bending stress (developed at the time of passage through the pulley) to the corroded outer element wire, resulting finally in its breakage.

The stresses induced in the wire rope are the tensile stress of 69 kgf/mm² developed by tension (20 tf) and the bending stress of 32 kgf/mm² which is developed at the time of passage through the pulley, the total being about 100 kgf/mm² maximum. Judging from the values given in Table 4.5, therefore it is not unreasonable to consider that a severely corroded element wire is broken immediately.

1. Tension $P = 20\,\text{tf} = 20\,000\,\text{kgf}$
2. Cross-sectional area A of wire rope $= 289.5\,\text{mm}^2$
3. Tensile stress $\sigma_T = P/A = 69\ \text{kgf/mm}^2$
4. Modulus of elasticity of wire rope, E_r: 9000 kgf/mm²
5. Pulley diameter $D = 450\,\text{mm}$
6. Diameter of element wire, $\delta = 1.6\,\text{mm}$
7. Bending stress developed at the time of passage through pulley, $\sigma_b = E_r\ \delta/D = 32\ \text{kgf/mm}^2$

(iv) Conclusion and countermeasures

The failure of the wire rope is not attributable to defects in the properties of the material but is attributed to the corrosion caused by frequent exposure to seawater during inboard storage (on deck) for four years and five months. In other words, as the wire rope is built of many element wires laid together, the deterioration in rope strength resulting from local corrosion loss should be taken into due consideration even if the wire rope looks new. Namely, the strength of wire rope is the sum of the minimum strength of element wires in two or three pitches. To prevent such a break, complete corrosion preventive means should be taken when storing wire rope.

Table 4.6   *Summary of failures of wire ropes*

Description	Case 1 Break from equalizer wheel	Case 2 Break from the portion passing through the sheave	Case 3 Break from the end stopper of wire rope	Case 4 Break in nearly new wire rope
Breaking load of wire rope (capacity)	34.7 tf (30 tf)	7.6 tf (5 tf)	5.0 tf (2.8 tf)	39.3 tf Rating 15 tf, max 20 tf)
Life of wire rope to break (past example)	11 months (13 and 9 months)	8 months (9, 8 and 10 months)	17 months (about 50 months)	5 hours (53 months)
Lifting load at the time of break	16.9 tf	3.0 tf	3.4 tf	20 tf
Broken position	Equalizer wheel	Portion passing through the sheave	End stopper of wire rope	Portion passing through the sheave
Main case of break	Fatigue fracture due to wear and fatigue repeated slip	Fatigue fracture due to wear and fatigue caused by repeated use	Fatigue fracture from end stopper caused by excessive load developed by the washer placed wrong side up	Deterioration of strength due to corrosion loss resulting from the exposure to sea-water during inboard storage for 4 years and 5 months

4.3.3 Summary

The results described above may be summarized as shown in Table 4.6. It will be noted from these examples that the wire rope is designed on the basis of 'damage tolerance design' and its service life gradually decreases after the start of use. It is therefore considered that the wire rope will eventually break at some time. The precaution 'Do not come under a lifted load' is quite a reasonable one for the protection of workers.

References

1. Nishida, S., Urashima, C. and Masumoto, H. (1982) Preprints for the Fractography Committee, The Society of Materials Science, Japan
2. Nishida, S. (1982) *Journal of the Japan Society of Mechanical Engineers*, **87**, No. 776
3. Wire Rope Handbook Editorial Committee (1967) *Wire Rope Handbook*, Hakua Shobo, Tokyo, p. 736
4. Nagano, S. (1968) *Journal of Metallic Materials*, **8**, 61, Tokyo
5. Nishioka, T. (1976) *Journal of the Japanese Society for Strength and Fracture of Materials*, **10**, 137
6. Seki, M., Yamamoto, S., Shinpo, T. and Toyokawa, T. (1969) *Journal of the Society of Materials Science*, **18**, 590
7. Doi, A. and Kawabata, Y. (1975) *Fatigue Symposium of JSMS*, Preprint 95

4.4 Failure of transmission shaft

As shown in Figure 4.35, failure of the shaft is second in frequency to failure of the weld zone in failures of equipment and machine parts. Failure of the transmission shaft in particular results in immediate stoppage of the machine, because the transmission shaft is designed to transmit the power. Unlike the bearings, the transmission shaft is not standardized. For this reason, spare shafts are not on hand at almost all plants. If the transmission shaft fails, therefore, a long period is required for its repair.

4.4.1 Failure of the travelling shaft of a crane

(a) Outline of failure
The travelling shaft of a crane (rating: 35 tf) and the position of failure are shown schematically in Figure 4.35, while the fracture surface is shown in Figure 4.36. The shaft is made of S35C. The fracture surface is nearly vertical to the axis of the shaft. The main point of failure initiation is the double notched part, i.e. the rounded part ($R = 3$ mm) of the stepped shaft between the key way ($\phi 90$ mm) and the bearing ($\phi 100$ mm) of the travelling shaft coupling. The failure of the travelling shaft paralyses the plant operation, with a serious effect on production.

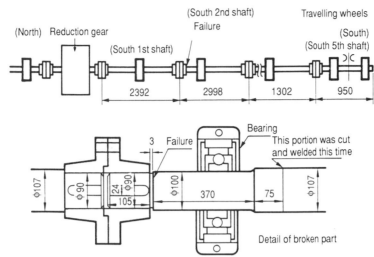

Figure 4.35 Position of failure

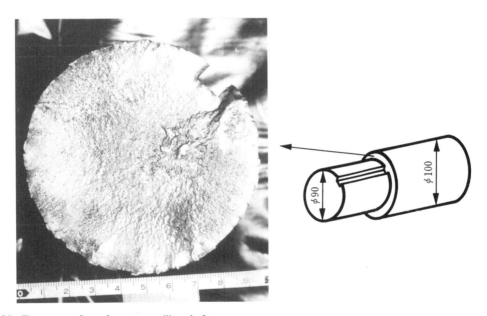

Figure 4.36 Fracture surface of crane travelling shaft

(b) Investigation and analysis of the cause of failure

(i) Macroscopic observation
From the results of macroscopic observation of the fracture surface shown in Figure 4.36, the failure of the shaft is judged to be a fatigue failure caused by repetition of torsional stress. The reasons for the decision are as follows.

1. The fracture surface consists of a comparatively smooth peripheral part and a central part resembling a chrysanthemum. The former is the fatigue failure surface and the latter is the final failure surface due to torsion.

2. The main point of failure initiation is estimated to be the point where the key way meets the stepped portion. Many secondary initiation points are observed in the rounded part of the stepped shaft.

3. The smooth fatigue fracture surface is 30 mm in depth and the final fatigue fracture surface is also about 30 mm in depth.

(ii) Hardness distribution and microstructure
To ascertain whether fatigue failure is attributable to

defects in the properties of the material, the micro-structure and hardness in the vicinity of the failure initiation point were investigated (micrographs are omitted).

The distribution of hardness in the cross section of the broken shaft is shown in Figure 4.37. The surface of the shaft was subjected to a welding build-up of thickness 2–3 mm. The hardness H_v of the welding build-up zone is about 200, which is slightly higher than that of the base metal (S35C). The base metal is of ferrite–pearlite structure, but the weld zone is of bainite structure. Our observation did not reveal defects, such as weld cracks. The question whether the welding build-up has an adverse effect on the failure of the shaft needs to be studied, and the results obtained by Takahashi, Takashima and Ito [1] are useful. They conducted a plane bending fatigue test of notched specimens of welded joints and base metals for SM50, WT60 and WT80, with a stress concentration factor $\alpha = 2$. The results of the test are shown in Table 4.7 and Figure 4.38. For all steel grades, the notch fatigue strength of the respective structures of the welded joints is higher than for base metals, suggesting that the heat of the welding build-up does not have an adverse effect on the fatigue strength. The load applied to the broken shaft is mainly a torsional load, while the results obtained by Takahashi, Takashima and Ito are the results for plane bending. However, as different types of loading will not have a different effect on the fatigue strength of the structure, it is judged that welding build-up has no adverse effect on shaft failure.

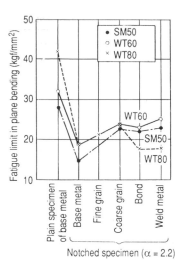

Figure 4.38 Fatigue limits of specimens of base metals and welds in notched plane bending

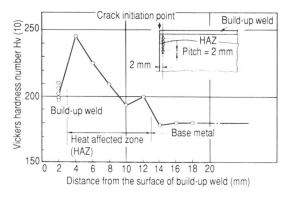

Figure 4.37 Macroscopic illustration of cross section of the broken shaft and hardness distribution

(iii) *Estimation of the torsional fatigue strength of a broken shaft*

The torsional fatigue limit of a plain specimen of S35C ranges from 11 to 20 kgf/mm², normally 13–14 kgf/mm². As the hardness of the broken shaft is about 180, the torsional fatigue limit of the shaft is estimated to be 15 kgf/mm². As this value is for the standard test specimen, the fatigue limit of the broken shaft itself is

estimated as described below. That is, the decrease in the fatigue limit due to (a) the size effect, (b) surface finish, and (c) a notch is estimated.

We consider first the decrease in fatigue strength due to the size effect. It is well known that fatigue strength expressed in relation to stress decreases with increasing size of the member. This is called the size effect [2]. One of the reasons for the decrease in fatigue strength is that the stress gradient decreases with increasing size (Figure 4.39). Moreover, if the surface area of volume is increased, the probability that large defects will exist in a member is increased. Furthermore, residual stress is more likely to be created as the size increases, and the stress in the cross section is not necessarily uniform, although it cannot be included in the size effect. Because of the effects described above, the fatigue strength of members having a far larger size than the

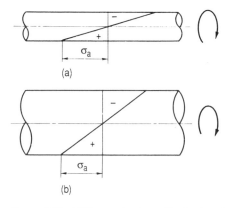

Figure 4.39 Difference in stress distribution due to difference in size: (a) small diameter; (b) large diameter

Table 4.7   *Hardness distribution in base metal and structure of welds*

Kind of steel	Base metal (M)	Fine grain (F)	Coarse grain (C)	Bond (B)	Weld metal (D)
SM50	Ferrite + pearlite, H_v 155–160	Fine ferrite + pearlite, H_v 175–180	Bainite, H_v 230–240	H_v 170–175	Ferrite + bainite, H_v 200–210
WT60	Tempered bainite, H_v 190–200	Transitional structure, H_v 190–200	Bainite, H_v 270–280		H_v 245–250
WT80	Tempered martensite, H_v 245–260	Transitional structure, H_v 230–240	Martensite + bainite, H_v 330–340		Ferrite + bainite

specimen becomes lower than that of the standard specimen. However, quantitative evaluation of the latter two factors is difficult.

Accordingly, the decrease in fatigue strength due to the size effect is calculated below. The coefficient of size effect is given by the following equation [2]. For rotational bending fatigue:

$$\zeta_b = \frac{\sigma_{\omega bd}}{\sigma_{\omega b10}} = 1 - \frac{\sigma_{\omega b10}}{\sigma_B}(0.522\ e^{-5.33/d} - 0.306)$$

$$(4.13)$$

where $\sigma_{\omega b10}$ is the fatigue limit of a plain specimen 10 mm in diameter (kgf/mm^2), σ_B is the tensile strength (kgf/mm^2) and d is the diameter (mm).

For torsional fatigue, the following equation is given based on the strain energy theory for rotational bending fatigue:

$$\zeta_t = \frac{\tau_{\omega bd}}{\tau_{\omega 10}} = 1 - \sqrt{3}\ \frac{\tau_{\omega 10}}{\sigma_B}(0.522\ e^{-5.33/d} - 0.306)$$

$$(4.14)$$

As the equations shown above are experimental equations, it is desirable to allow a safety factor of 1.1 for rotational bending fatigue and of 1.2 for torsional fatigue for the values calculated by these equations.

Since the shaft is 90 mm in diameter, it is considered that the stress applied to the broken shaft is mainly torsional stress. Accordingly, the coefficient of size

effect, ζ_t, is calculated as shown below from equation (4.14):

$$\zeta_t = 1 - \sqrt{3} \times \frac{15}{60}(0.522\ e^{-5.33/90} - 0.306) = 0.919$$

$$(4.15)$$

For reference, an example of the effect of diameter on repeated fatigue limit is shown in Table 4.8 [3].

We consider next the effect of surface finish on fatigue strength. Even visual observation reveals that the surface finish of the rounded part of the step of the broken shaft is very rough. The effect of surface finish on fatigue limit is shown in Table 4.9 [3]. In the case of the broken shaft, the surface finish is judged to be rough turning (V). Accordingly, the fatigue limit reduction factor ζ_f due to surface finish is estimated to be about 0.84 [4].

Table 4.8   *Effect of diameter on torsional fatigue limit* [3] *Ni-Cr-W steel for crank shaft**

	Fatigue limit (kgf/mm^2)		
Shape of shaft	$d = 14$ mm	$d = 30$ mm	$d = 45$ mm
Plain, solid	28	26	20
Plain, hollow	24	21	18
Plain, with rim	25.5	–	19
With key way	17	14.5	13
With side hole	16	13	11.5

* Quenched in oil from 820°C after forging, tempered at 580°C, $R_m = 92$ kgf/mm^2 (R_m is the tensile strength).

Table 4.9   *Effect of finishing on fatigue limit*

	Fatigue limit			
	0.49% C steel		0.02% C steel	
Surface condition	(kgf/mm^2)	(%)	(kgf/mm^2)	(%)
Emery paper no. 0, 00	35.2	100	18.5	100
Emery paper no. 0, 00 and emery paper no. 1, 0, 00 followed by iron oxide red buff finishing	36	103	–	–
Cylindrical grinding	32.3	92	–	–
Finish turning	30.7	87	17.1	92
Rough turning	29.6	84	16.5	88

Table 4.10   *Torsional fatigue limit of stepped round bar* [3]

| Kind of steel | Static mechanical properties | | | | Diameter of stepped round bar | | | | | Fatigue limit | | | | | |
	R_e(kgf/mm²)	R_m(kgf/mm²)	A (%)	S (%)	D (mm)	d (mm)	r (mm)	d/D	r/d	σ_D (kgf/mm²)	K_f	K_t (Lehr)	q	κ_t (Thum)	q
StC35 61 Quenched and tempered	42.5	62.8	19.4	69.8	12.5	12.5	–	1	–	22.4	1	–	–	–	–
					17.9	12.5	3	0.7	0.24	20.8	1.09	1.42	0.21	(1.3)	0.30
					17.9	12.5	1	0.7	0.08	12.1	1.24	2.0	0.24	1.62	0.39
					17.9	12.5	0	0.7	0	11.8	1.90	(5)	0.23	(4.0)	0.30

K_f, notch factor; q, notch sensitivity; K_t, stress concentration factor; R_e, yield strength; R_m, tensile strength; A, elongation; S, reduction of area; estimated value in parentheses.

Finally, we consider the decrease in fatigue strength due to notch. The broken shaft has the so-called double notch in which the key way extends up to the step of the round bar. In the case of the double notch, the notch factor β (=fatigue limit of plain specimen/fatigue limit of notched specimen) is calculated by multiplying the notch factors of respective notched specimens ($\beta = \beta_1\beta_2$) [5]. In other words, the notch factor β_1 of the stepped round bar and the notch factor β_2 of the round bar with key way are calculated separately and the notch factor β of the broken shaft is obtained by multiplying β_1 by β_2.

The notch factor β_2 of the stepped round bar was obtained from Table 4.10. Although the ratio d/D shown in Table 4.10 differs slightly from that of the broken shaft, the notch factor β_1 of the stepped round bar can be estimated as about 1.85 as the ratios d/D and r/d of the round bar are 0.9 and 0.03, respectively. The notch factor of the round bar with key way was obtained from Table 4.8. Although the steel grade of the broken shaft differs from that shown in Table 4.8, the notch factor β_2 of the round bar with key way can be estimated as $28/17 = 1.65$ from the '$d = 14\,\text{mm}$' column. Accordingly, the notch factor β of the broken shaft is $\beta_1\beta_2 = 1.85 \times 1.65 = 3.05$.

Taking these factors into consideration, the torsional fatigue limit τ_w of the broken shaft is estimated as shown below:

$$\tau_\omega = \frac{\tau_{\omega 0}\,\zeta_t\zeta_f}{\beta} = \frac{15 \times 0.92 \times 0.84}{3.05} = 3.8 \text{ kgf/mm}^2$$

(4.16)

where $\tau_{\omega 0}$ is the torsional fatigue limit of a plain specimen, ζ_t is the fatigue strength reduction factor due to the size effect and ζ_f is the fatigue strength reduction factor due to surface finish. In short, the torsional fatigue limit of the broken shaft is 3.8 kgf/mm².

(iv) Estimation of stress actually applied to the broken shaft and the number of cycles to failure

According to a calculation made by the mechanical engineering department on site [6], the stress τ applied to the broken shaft is 2.84 kgf/mm² in the case of rated torque. This is the stress to be applied when the crane travels under rated torque. Maximum torsional stress is created when the crane is started and stopped. It is said that maximum torque is usually about 1.5 times rated torque.

$$\tau_{max} = 1.5\tau = 1.5 \times 2.84 = 4.3 \text{ kgf/mm}^2 \quad (4.17)$$

In other words, the maximum torsional stress developed in the broken shaft is estimated to be 4.3 kgf/mm².

On the basis of the results described above, an estimated S–N curve for the broken shaft is shown in Figure 4.40 in comparison with the actually applied

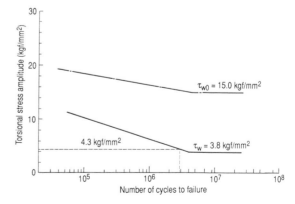

Figure 4.40   Torsional S–N curves for the broken shaft (estimated)

stress. If the application of torsional stress is repeated with an amplitude of 4.3 kgf/mm², the shaft will break after cyclic stressing of about 290×10^4 cycles. According to a calculation made by the mechanical engineering department on site, the number of cycles to failure, N_f, is 252×10^4. The order of both values is the same.

(c) Summary and countermeasures

The failure of the shaft is attributable to fatigue caused by repeated stressing. The following three causes can be pointed out as major causes.

1. In the shaft, the key way extends up to the rounded part of the stepped shaft.
2. The radius of the notch in the rounded part of the stepped shaft is small.
3. The surface finish is rough.

Cause 1 in particular decreases the fatigue strength markedly because of the multiplier effect of stress concentration at the key way and stress concentration at the rounded part of the stepped shaft. According to Yoshitake [7], the notch factor for the case where the key way cuts into the rounded part of the stepped shaft is greater than for the case where the key way is separated from the rounded part of the stepped shaft, but smaller than the product of both factors. However, the factors for the case where the key way extends up to the rounded part of the stepped shaft are not available. To ensure safety, therefore, a calculation using the multiple effect is considered better.

Taking these causes into consideration, the following countermeasures are considered effective.

1. The key way and the rounded part of the stepped shaft should be separated from each other by a distance corresponding to at least the height of the step or the width of the key way, whichever is larger. Because of the separation, the notch factor β is decreased to either β_1 or β_2.
2. The radius of the rounded part of the stepped shaft is increased. For example, the torsional fatigue limit is improved by about 10% by increasing this radius from 3 to 5 mm, and fatigue failure will therefore be eliminated.
3. The rounded part of the stepped shaft is finished by using No. 400 emery paper. This finish improves the torsional fatigue limit by about 20%.

4.4.2 Failure of the counter shaft of a water pump

The position of failure of the counter shaft of a water pump is shown schematically in Figure 4.41. The shaft is made of S45C. The appearance of the broken counter shaft is shown in Figure 4.42. All traces of the failure have disappeared because the fracture surfaces rub against each other. If the revolving shaft is broken, the fracture surfaces rub against each other, leaving no

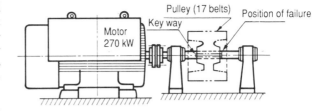

Figure 4.41 Outline of broken portion of counter shaft of water pump

(a)

(b)

(c)

Figure 4.42 Failure of counter shaft of water pump: (a) appearance of broken shaft; (b) broken portion; (c) enlarged view of (a)

trace of the failure on the fracture surfaces in many cases. For this reason, fractography cannot show its effect. The failure initiates at the double-notched part between the key way and the rounded part ($R = 1$ mm) of the stepped shaft in the same way as described in Section 4.4.1. The failure of the shaft is attributable to the fatigue caused by repetition of torsional stress on starting and stopping. In the case of the broken shaft, it is estimated that a torsional stress of 8.5–9.5 kgf/mm^2 was repeatedly applied. As the diameter of the shaft is 85 mm, the torsional fatigue limit is estimated to be about 18 kgf/mm^2 according to the results of a test on a standard specimen. A description of the method of analysis is omitted because it is the same as that given in Section 4.4.1.

The double notch design as described above should be absolutely avoided as it decreases the fatigue strength markedly. Otherwise, it is necessary to allow for a very high safety factor.

References

1. Takahashi, K., Takashima, H. and Ito, A. (1967) *Journal of the Iron and Steel Institute of Japan*, **53**, 518, Tokyo
2. Nakamura, H., Tsunenari, T., Horikawa, T. and Okazaki, S. (1983) *Failure Life Design of Machines*, Yokendo, Tokyo, p. 104
3. Cazaud, R. *et al.* (1973) (translated by S. Nisijima and H. Funakubo) *Fatigue of Metals*, Maruzen, Tokyo, pp. 337 and 318
4. Ishibashi, T. (1977) *Fatigue of Metals and Prevention of Fracture*, Yokendo, Tokyo
5. Kawamoto, M. (1962) *Fatigue of Metals*, Asakura Shoten, Tokyo, p. 108
6. Nemoto, Y. (1977) *NSC's Technical Paper*, 1, Nippon Steel Corporation
7. Yoshitake, H. (1978) *Handbook for Fatigue Design of Metal Materials*, (ed. JSMS) Yokendo, Tokyo, p. 59

4.5 Failure of crank of an overcurrent breaker

The crank of an overcurrent breaker (OCB) connected to the 3000 kW motor of a rolling mill broke, resulting in a halt in operations of about 30 hours, and large losses were thereby incurred.

4.5.1 Outline of failure

The OCB is located between the transformer and the rolling mill motor. This OCB is designed to automatically shut off the circuit to protect the motor when the motor is overloaded. The OCB and the position of

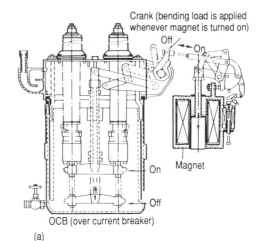

(a)

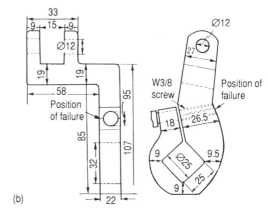

(b)

Figure 4.43   (a) Outline of OCB and (b) position of failure of crank

failure are shown schematically in Figure 4.43. The failure initiates at the crank which connects the magnet of the OCB to the circuit breaker. It is known that the OCB is subjected to about 7000 repetitions of load per year. Figure 4.44 shows the position of failure in the crank and the macroscopic fracture surface. The failure initiated at the end of the tapped hole for the bolt which is used for fastening the crank shaft (the part which is seen as blackish in Figure 4.44(b): in many cases, the part where cracks arise in the early stages becomes darker in colour than other parts). The OCB had been in service for about 13 years.

4.5.2 Investigation and analysis of the cause of failure

(a) Microstructure
The microstructure in the vicinity of the broken point is shown in Figure 4.45. The crank is made of black heart malleable cast iron equivalent to FCMB-32. No abnormalities are detected in the structure.

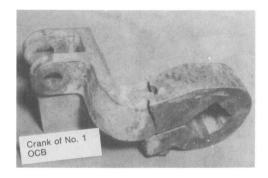

Figure 4.44 Position of failure of crank and fracture surface

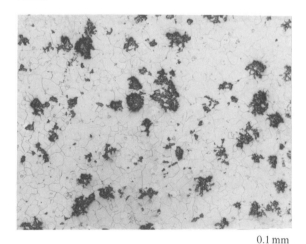

0.1 mm

Figure 4.45 Microstructure in the vicinity of the position of failure

(b) Observation of fracture surface

According to the results of macroscopic and microscopic observations, the fracture surface at the edge of the tapped hole for the crank shaft clamp bolt is the oldest surface, indicating that this part is the failure initiation point. Figure 4.46 shows a sketch of the fracture surface and an example of striation observed

at part A. The failure of part A is attributable to fatigue. Although photographs are omitted, observation of part B of Figure 4.46 revealed quasi-cleavage and dimple fracture surfaces. It is therefore apparent that part B is the final failure surface and that the failure occurred suddenly.

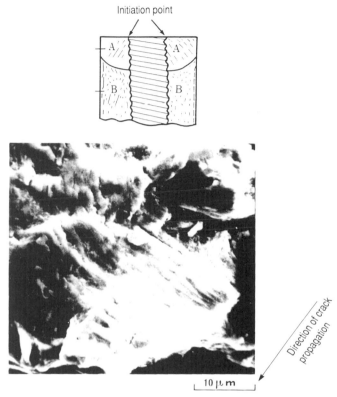

Figure 4.46 Striation observed at part A

(c) Measurement of stress and analysis of failure

To investigate the cause of failure of the OCB crank, the stress applied to the crank was measured by attaching a strain gauge in the vicinity of the broken part which is shown in Figure 4.43. The results of measurement are shown in Figure 4.47. When the switch is opened and closed, the strain (stress) wave form is changed from that observed during operation.

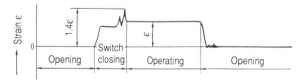

Figure 4.47 Strain waveform for closing, operation and opening

When the switch is closed, a strain (stress) which is about 1.4 times the strain generated during operation is momentarily applied. The repeated bending stress applied to the broken part is calculated on the basis of the results shown in Figure 4.47.

As the crank is made of a material equivalent to FCMB32 ($\sigma_B \geq 32$ kgf/mm^2, $\sigma_{0.2} \geq 19$ kgf/mm^2 according to the Japanese Industrial Standard (JIS) [1], the actual tensile strength is estimated to be about 35 kgf/mm^2. In this case, the Young's modulus E_c will be 13.0×10^4 kgf/mm^2 [2] and the nominal stress will be 0–17.3 kgf/mm^2. The fatigue strength of a plain specimen in rotational bending is assumed to be about 15.0 kgf/mm^2 (fatigue limit) [3]. Accordingly, σ_a is calculated as 10.5 kgf/mm^2 from the corrected Goodman equation:

$$\sigma_a = \sigma_w (1 - \sigma_m/\sigma_B) \qquad (4.18)$$

where σ_w is the completely reversed fatigue limit, σ_a is the fatigue limit (in the stress range) under the mean stress σ_m and σ_B is the tensile strength. In other words, the fatigue limit (in the stress range) σ_{u0} is 21.0 kgf/mm^2 in the case of completely tensile plane bending. According to other literature [4], the bending fatigue limit (in the case of completely tensile plane bending stress) is 22.5 kgf/mm^2 when the tensile strength is 33.8 kgf/mm^2. Accordingly, the bending fatigue limit in completely tensile plane bending, σ_{u0}, is taken as 22.0 kgf/mm^2.

Thus, the stress concentration factor α at the broken part is calculated as 2.25 [5]. Cast iron is unexpectedly high in fatigue strength and the effect of a notch on this strength is small. In the case of carbon steel, the stress concentration factor α is nearly equal to β (β is the notch factor = fatigue limit of plain specimen/fatigue limit of notched specimen) when α is equal to or smaller than 2.3, but α becomes larger than β when α is

larger than 2.3 [6]. As the member being discussed here has mill scale, it is judged that a notch factor β of 2 (this is the factor for black heart malleable cast iron) will suffice. That is, the fatigue limit σ_u of a notched specimen under completely tensile plane bending stress is calculated as 11.0 kgf/mm^2. The above description is illustrated by the S–N curve in Figure 4.48.

The number of cycles to failure, N_f, can be obtained by drawing a line corresponding to the repeated stress applied to the crank, $\sigma_{ua} = 17.3$ kgf/mm^2, to the S–N curve (solid line) as shown in Figure 4.48. That is, the number is 9.2×10^4. This value is very close to the number of cycles applied to the broken crank, which is about 9.1×10^4.

4.5.3 *Summary and countermeasures*

The failure of the OCB crank is not attributable to defects in the properties of the material but is attributed to the fatigue failure caused by repeated application of impulsive bending stress to the crank due to the opening and closing of the OCB during a long period of time.

As a countermeasure, the OCB crank should be so designed as to prevent fatigue fracture. That is, one of the following three methods should be adopted.

1. A tapped hole about 15 mm in depth is provided for the W3/8 bolt instead of a through hole.
2. Nodular cast iron with a tensile strength $\sigma_B > 50$ kgf/mm^2 is used.
3. If the material now in use is not changed, the thickness of the broken part should be increased by more than 1.25 times the current thickness (1.60 times in the case of the width).

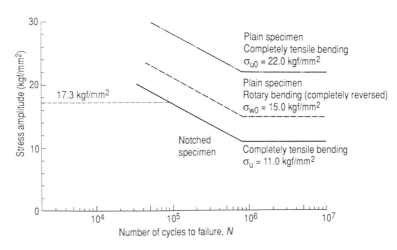

Figure 4.48   *S–N curve for black heart malleable cast iron*

References

1. Japanese Industrial Standard, JIS G 5702
2. The Japanese Association of Casting Industry (ed.) (1975) *Properties of Cast Iron*, Coronasha, Tokyo, p. 104
3. The Japan Society of Mechanical Engineers (ed.) (1974) *Design Data for Fatigue Strength of Metals*, p. 76
4. The Japanese Association of Casting Industry (ed.) (1975) *Properties of Cast Iron*, Coronasha, Tokyo, p. 105
5. Nishida, M. (1972) *Stress Concentration*, Morikita Shuppan, Tokyo, p. 309
6. Ishibashi, T. (1954) *Fatigue of Metals and Prevention of Fracture*, Yokendo, Tokyo, p. 55

4.6 Failure of discharge wires of an electrostatic space cleaner (ESC)

4.6.1 Outline of failure

Failure of the star-shaped discharge wires used in an electrostatic space cleaner (ESC) at a sintering plant was detected. As the replacement of the broken wire is not easy and many wires are used in the ESC, the cause of failure was investigated in detail and the possibility of a future increase in this type of failure was studied to work out rigorous countermeasures.

1. The structure of an ESC is shown in Figures 4.49–4.51. The ESC is hammered three times per hour to remove the dust adhering to the star-shaped discharge wires in the frame, which is 12 m high, 7 m wide and 26 m long.
2. The atmospheric temperature in the ESC is normally 120–140°C, the maximum temperature being 200°C.
3. The ESC had been in service for one year, six months and six days before failure of the discharge wires was detected at the time of an interior inspection.
4. The number of broken star-shaped-type discharge wires is five. (The total number of discharge wires used in the ESC is 6000, the length of each wire being 5000 mm.)
5. The steel grade is SUS304.
6. The length of the diagonal line of the wire is 6 mm. The wire has a star-shaped cross section.

4.6.2 Investigation of the cause of failure

(a) Items investigated

1. Chemical analysis
2. Observation of fracture surface (macroscopic and microscopic)
3. Microstructure
4. Hardness distribution

(a)

(b)

Figure 4.49 General view of ESC: (a) broken ESC system; (b) ESC system under construction in adjacent area

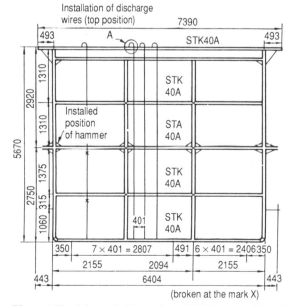

(broken at the mark X)

Figure 4.50 Schematic illustration of ESC and position of failure

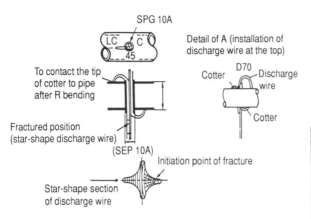

Figure 4.51 Installation of star-shaped discharge wire and its condition of failure

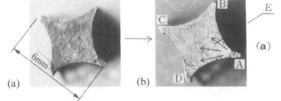

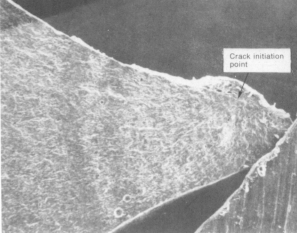

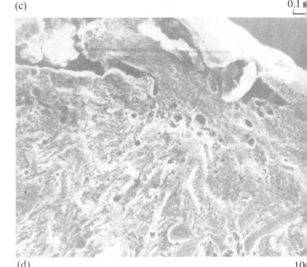

(b) Results of analysis

(i) Chemical analysis

The chemical composition is shown in Table 4.11. The composition meets the Japanese Industrial Standard (JIS) for SUS304.

(ii) Observation of fracture surface

The results of observation of the fracture surface by SEM are shown in Figures 4.52–4.54. The whole specimen, including the fracture surface, was covered with dust which is assumed to be sinter dust. When the dust was removed by pickling, the initiation and propagation of cracks could be clearly observed. Cracks initiated in the vicinity of the vertex of the star-shaped cross section and propagated in the diagonal direction, resulting in final fracture. In the fracture surface, striation peculiar to fatigue was observed. It is therefore considered that the failure of the discharge wire is fatigue failure caused by the impact force repeatedly applied to remove sinter dust.

In the vicinity of the crack initiation point, a small burr is formed on the surface. It is estimated that the burr was formed when the corner of the discharge wire was struck in error at the time of fastening the discharge wires with cotters (see Figure 4.51). The fatigue strength varies, depending on the balance between the surface effect of the burr, work hardening and compressive residual stress. As a fatigue crack initiated at the burr, it is judged that the fatigue strength of the wire was decreased by the burr. However, the effect of the burr is small.

Figure 4.52 Results of observation of fracture surface by SEM: macroscopic fracture surface (a) before and (b) after pickling; (c) vicinity of point A; (d) enlarged view of crack initiation point

Table 4.11 *Chemical composition* (wt %)

C	Si	Mn	P	S	Ni	Cr	Mo	Nb	Ti	Remarks
<0.080	<1.00	<2.00	<0.20	<0.060	8.00/10.50	18.00/20.00	–	–	–	JIS standards
0.079	0.55	1.26	0.018	0.005	9.30	18.75	0.088	0.001	0.001	

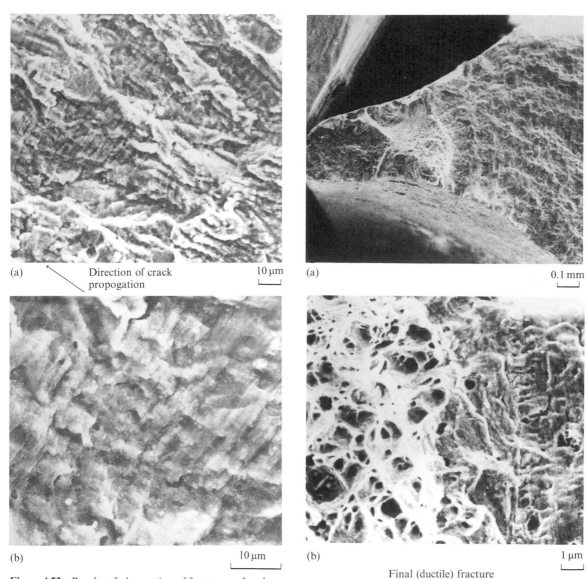

(a)   Direction of crack propogation   10 μm

(b)   10 μm

(a)   0.1 mm

(b)   1 μm

Final (ductile) fracture

Figure 4.53 Results of observation of fracture surface by SEM (point E in Figure 4.52): (a) striation at point E (2.1 mm from the vertex); (b) enlarged view of (a)

Figure 4.54 Results of observation of fracture surface by SEM (point C in Figure 4.52): (a) final fractured part (point C); (b) boundary between fatigue and final (ductile) fracture

(iii) Microstructure

The optical micrograph of a star-shaped discharge wire is shown in Figure 4.55. The structure is a typical austenitic stainless steel structure which was completely solution-treated. The grain size number is 5.

(iv) Hardness distribution

The hardness distribution in the cross section of a star-shaped discharge wire is shown in Figure 4.56. H_v (10 kgf) ranges from 182 to 185, satisfying the JIS requirements (below 200). When the measurement was made under a light load (100 gf), however, the hardness was relatively high. The hardness at the surface was higher by about 20% than that in the centre.

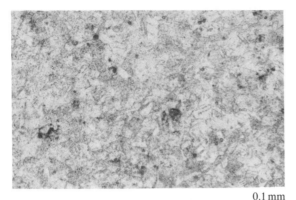

0.1 mm

Figure 4.55   Optical microstructure of SUS304

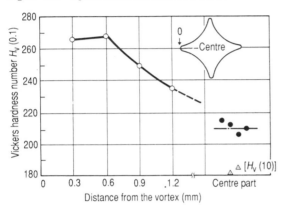

Figure 4.56   Hardness distribution in the vicinity of vertex and at the centre

4.6.3   *Estimation of approaching failure by stress measurement*

(a)   Number of cycles to failure

1. Duration of use: about 18 months
2. Frequency of hammering during 18 months:
   $N = 18$ months $\times$ 30 days $\times$ 24 hours $\times$ 3 times/hour
   $= 3.9 \times 10^4$ cycles

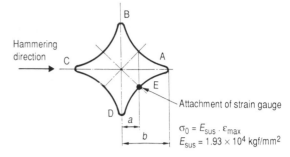

Figure 4.57   Stress measurement

(b)   Preconditions

1. σ_a: stress amplitude, converted from the results of stress measurement (see Figure 4.57)
2. Measuring condition: strain gauge KFC-5-C1-11, gauge length: 5 mm, 12 points
3. Dynamic strain gauge: manufactured by Kyowa Dengyo Co. Ltd, 0–1 kHz
4. Electromagnetic oscillograph: manufactured by Yokogawa Hokushin Electric Corporation, 18 points, 5 cm/min–200 cm/sec
   Strain → Dynamic → Electromagnetic → Data
   gauge   strain   oscillograph   analysis
             gauge

                                    Frequency
                                    analysis

In Figure 4.57, the stress at the point E where the strain gauge was set was 10.2 kgf/mm^2. The stress is converted into that at the point A as shown below.

$$\sigma_a = 10.2 \times \frac{b}{a} = 10.2 \times \frac{19}{7} = 27.7 \text{ kgf/mm}^2$$

(proportional calculation for enlarged size).
N is the number of cycles to failure. With this type of structure, vibration does not attenuate markedly. From the results of measurements, it is considered that N is about ten times the number of hammering cycles, which is 3.9×10^4 cycles (because of vibration). For the calculation, N is taken as 4×10^5. Temperature: 200°C.

Table 4.12   *Mechanical properties (room temperature–high temperature)**

Kind of steel	Test temperature	$\sigma_{0.2}$ (kgf/mm^2)	σ_B (kgf/mm^2)	El (%)	ϕ (%)	Bending test (RT)
YUS304 N	RT	49.3	82.9	48.6	66.5	Contact bending ($R=0$ mm);
	200°C	31.8	65.4	46.7	66.1	no cracks; $t=12$ mm
	400°C	26.8	61.9	45.8	63.0	
	600°C	23.3	53.1	41.2	61.7	
SUS304	RT	24.0	59.5	65.0	75.5	Contact bending ($R=0$ mm);
	200°C	17.5	44.5	53.5	73.5	no cracks; $t=13$ mm
	400°C	15.4	40.5	45.0	70.0	
	600°C	14.0	38.0	43.8	73.0	

* See Figure 4.60. $\sigma_{0.2}$, 0.2% proof stress; σ_B, tensile strength; El, elongation; ϕ, reduction in area.

(c) Estimation of fatigue strength

The relation between the tensile strength σ_B and the fatigue limit σ_w under the conditions shown in (b) above is calculated from Tables 4.12 and 4.13 and Figure 4.58 [1]. As a result, a linear relation of $\sigma_w \fallingdotseq (1/2)\sigma_B$ is obtained (see Figure 4.59). The fatigue limit calculated from the tensile strength of SUS304 at 200°C is 22 kgf/mm². The S–N curve estimated from this fatigue limit is shown by the line with alternate long and short dashes in Figure 4.60. The fatigue strength σ_f which can sustain cyclic stressing of 4×10^5 cycles is 26.0 kgf/mm². As the repeated stress σ_a is $27.7 > \sigma_f = 26.0$ kgf/mm², the possibility that failures will occur is high.

Table 4.13 *Fatigue strength (cantilever rotary bending fatigue)**

Kind of steel	Fatigue limit (kgf/mm²)	Ratio of fatigue limit
YUS304 N	43	0.52
SUS304	29	0.49

* See Figure 4.60.

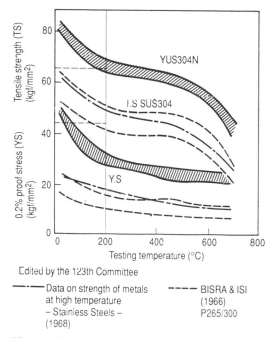

Edited by the 123th Committee

— ·— Data on strength of metals at high temperature – Stainless Steels – (1968)

– – – – BISRA & ISI (1966) P265/300

Figure 4.58 Comparison between SUS304 and YUS304N

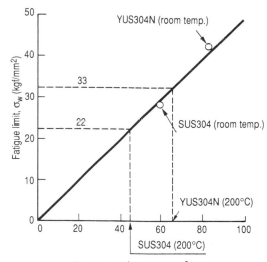

Figure 4.59 Relation between tensile strength σ_B and fatigue limit σ_w

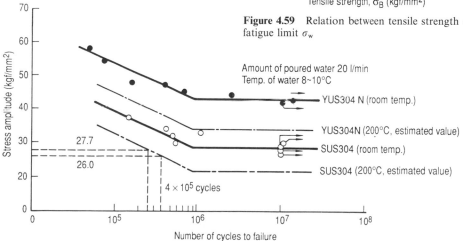

Figure 4.60 S–N curves for SUS304 and YUS304N (at room temperature and 200°C, respectively, cantilever rotary bending fatigue)

4.6.4 *Fracture mechanics approach and estimation of the number of cycles to failure*

(a) Analysis by simple assumption method

The propagation rate of fatigue crack in SUS304 at room temperature is calculated from the following equation [2]:

$$\frac{dl}{dN} = C(\Delta K_1)^m = 8.04 \times 10^{-11}(\Delta K_1)^{3.18} \quad (4.19)$$

As the stress intensity factor for the shape as shown in Figure 4.51 has not yet been calculated, the following assumptions are made to simplify the calculation. As shown in Figure 4.53, the fatigue crack propagated from point A to point C in the fracture surface. Accordingly, a two-dimensional assumption as shown in Figures 4.61 and 4.62 can be made. (The bending moment, cross-sectional area and specimen width in Figure 4.61(a) are assumed to be the same as those in Figure 4.61(b).)

According to the results of observation, the crack initiates at the burr. If the depth of the burr is assumed to be about 0.3 mm, the crack propagation in the second stage will be $l_i = 0.3$ mm to $l_c = 5.1$ mm. The stress intensity factor K_1 [3] is:

$$K_1 = \sigma_0 \sqrt{(\pi l)} \cdot F(l/w) \quad (4.20)$$

where nominal bending stress

$$\sigma_0 = 3PS/2W^2, \qquad \xi = l/W$$
$$F(\xi) = A_0 + A_1\xi + A_2\xi^2 + (l/W = \xi)$$

S is the span and W is the width of the test specimen (mm).

Since $S \gg W$, the following assumption can be made:

$$F(\xi) = A_0 = 1.12$$
$$K_1 = \sigma_0 \sqrt{(\pi l)} \cdot A_0,$$

As the cross section of the model is rectangular, the value 12.7 kgf/mm², which was obtained by conversion into the stress at point A in the rectangular cross section, is used as σ_0 instead of 27.7 kgf/mm² at point A in Figure 4.56.

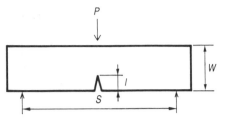

Figure 4.62   Loading condition of Figure 4.61(b)

Integrating equation (4.19):

$$dN = \frac{dl}{C(\Delta K_1)^m}$$

$$N = \int_{l_i}^{l_c} \frac{dl}{C(\Delta K_1)^m} = \int_{l_i}^{l_c} \frac{dl}{C(\sigma_0 A_0)^m(\pi l)^{m/2}}$$

$$= \frac{1}{C(\sigma_0 A_0)^m \pi^{m/2}} \int_{l_i}^{l_c} l^{-m/2} dl$$

$$= \frac{1}{C'} \frac{1}{(m/2)-1} \left[\frac{1}{l_i^{m/2-1}} - \frac{1}{l_c^{m/2-1}} \right] \quad (4.21)$$

where $C' = C(\sigma_0 A_0)^m \pi^{m/2}$

Substituting $C = 8.04 \times 10^{-11}$, $m = 3.18$, $\sigma_0 = 12.7$ kgf/mm², $A_0 = 1.12$, $l_i = 0.3$ mm and $l_c = 5.1$ mm into equation (4.21) and taking the number of cycles to crack initiation into account (1.2 times):

$$N_I = 14.5 \times 10^5 \text{ cycles} \quad (4.22)$$

The order of this N_I is the same as that of the number of cycles (4×10^5 cycles) calculated in Section 4.6.3.

(b) Determination of stress intensity factor by experiment

Let us determine the stress intensity factor experimentally according to the method proposed by Murakami *et al.* [4]. The material used for the experiment was a eutectoid steel which is less anisotropic than epoxy resin, is easy to process and exhibits brittleness at room temperature. The chemical composition of the

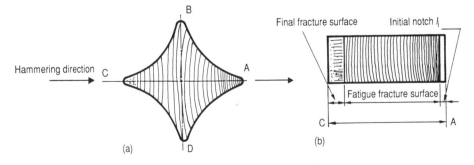

Figure 4.61   Model for calculation: (a) fracture surface; (b) model of (a)

material is shown in Table 4.14. All test specimens were taken from the rolling direction.

The shape of the specimen for the three-point bending test is shown in Figure 4.63. The specimen shown in Figure 4.63(a) is similar in cross section to the star-shaped discharge wire, while the specimen in Figure 4.63(b) has an ordinary rectangular cross section. Both specimens have the same width and

Table 4.14   *Chemical composition of eutectoid steel (wt %)*

C	Si	Mn	P	S	Al
0.68	0.22	0.96	0.017	0.014	0.018

Table 4.15   *Test results of three-point bending*

Notch depth (mm)	2	5	9	13	Mean value
σ_0/σ'_0	0.591	0.501	0.475	0.569	0.534

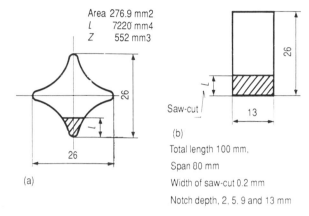

Area 276.9 mm2
l 7220 mm4
Z 552 mm3

26

26

(a)

Saw-cut l 13

(b)

Total length 100 mm,

Span 80 mm

Width of saw-cut 0.2 mm

Notch depth, 2, 5, 9 and 13 mm

Figure 4.63   Specimens for three-point bending test: (a) star-shaped section; (b) rectangular section. Total length, 100 mm; span, 80 mm; l, notch depth, four types (2, 5, 9 and 13 mm); width of saw-cut, 0.2 mm

notch depth. The stress intensity factors K_I' and K_I for both specimens are given by:

$$K_I' = \sigma_0' \sqrt{(\pi l)} \cdot F'(l/W) \tag{4.23}$$
$$K_I = \sigma_0 \sqrt{(\pi l)} \cdot F(l/W) \tag{4.24}$$

In equations (4.23) and (4.24), $F'(l/W)$ only is unknown. As K_{IC}' can be considered to be equal to K_{IC}:

$$F'(l/W) = \frac{\sigma_0}{\sigma_0'} F(l/W) \tag{4.25}$$

The results of the three-point bending fracture test are shown in Table 4.15 and Figure 4.64. It is considered that the value of σ_0/σ_0' in Table 4.15 does

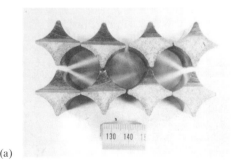

(a)

(b)

Figure 4.64   Results of three-point bending test (fracture surface): (a) star shaped; (b) rectangular shaped

not show a specific tendency even if the notch depth is increased. Accordingly, the mean value was used as the representative value of σ_0/σ_0'. That is, substituting $\sigma_0/\sigma_0' = 0.534$ into equation (4.20):

$$F'(l/W) = 0.534F(l/W) \qquad (4.26)$$

Substituting equation (4.26) into equation (4.20) integrating dN in the same manner as for equation (4.21), and substituting the numerical values, we obtain

$$N_2 = 14.9 \times 10^5 \text{ cycles} \qquad (4.27)$$

This value is nearly equal to N_1 calculated by equation (4.22) and N_2 calculated by equation (4.27).

4.6.5 *Summary and countermeasures*

From the results described above, it may be concluded that failure of the discharge wires of the ESC is a fatigue failure caused by hammering which is performed at regular intervals to remove the dust adhering to the discharge wires. The order of the number of cycles to failure by hammering (4×10^5) is the same as that of the number calculated on the basis of linear fracture mechanics (14.5×10^5 and 14.9×10^5). Even if the calculation is made by approximating the star-shaped cross section to the rectangular cross section (assuming that the cross-sectional area and bending moment are the same), approximate values which are usable in practice can be obtained.

For prevention of the failure, such measures as a decrease in the hammering force and a change in the wire supporting method are effective. A more effective method is to change the properties of the plain material now in use. That is, fatigue failure can be prevented by changing SUS304 now in use (fatigue limit $\sigma_{w0} = 29 \text{ kgf/mm}^2$) to YUS304N (fatigue limit $\sigma_{w0} = 43 \text{ kgf/mm}^2$).

Observation of the fracture surface revealed that the crack initiation point is the burr formed on the surface of the discharge wire. Special attention should therefore be directed against causing a sharp scratch or the like when the discharge wires are installed.

References

1. Nippon Steel Corporation (1975) Catalogue: Austenitic Stainless Steel with High Tensile Strength (YUS304N), p. 2
2. Urashima, C., Nishida, S. and Masumoto, H. (1981) Committee on Fatigue, Society of Materials Science, Japan, Preprint 11
3. Okamura, H. (1981) *Introduction to Linear Fracture Mechanics*, Baifukan, Tokyo, p. 218
4. Murakami, Y., Harada, S., Endo, T. *et al.* (1982) *Journal of the Society of Materials Science* (JSMS), **31**, 515, Kyoto, Japan

4.7 Failure of fastening screws (fatigue failure) [1–6]

A broad definition of fastening screws includes not only the combination of nuts and bolts but also fastened members. The representative fastener is the combination of a nut and a bolt. In general, fastening screws in a narrow sense have the following advantages.

1. They can be easily assembled and disassembled.
2. They can be set, while making necessary adjustments, or can be set with high precision with simple fastening tools.
3. As the wedge effect of threads can be utilized, even very thick members can be fastened tightly.

Because of these advantages, several thousand fastening screws are used, for example, in a car [7]. Moreover, surprisingly large numbers of bolts are used in a wide variety of machines and equipment, such as electrical equipment, machine tools, construction machinery, rolling-stock, steel towers, bridges, transportation equipment, etc.

Fastening screws are so widely used in daily life that their usefulness is not fully recognized. We are apt to attach less importance to screws by saying 'only one bolt', or 'only one screw', in spite of their important functions. The origin of screws dates back to the times before Christ in Europe but to the Meiji Era (1868–1911) in Japan [8]. In Japan, the fastening screw is one machine part the improvement of which does not attract wide attention. This may be because we have not had sufficient time for a full understanding of the functions of screws because of its short history in Japan or because we are apt to consider that there is no room for further improvement of screws because of their simple structure. In practice, machine design begins with major parts and ends with small parts, such as bolts. In many cases, therefore, the manufacture of bolts and other small parts is started without fully examining their safety, including fatigue. In short, bolts and the like are machine parts for which serious consideration is apt to be neglected.

The examples described below indicate the importance of fastening screws. On 4 February 1966, All Nippon Airways' regular flight from Chitose Airport in Hokkaido, in the northern part of Japan, crashed into the sea just before landing at Haneda Airport in Tokyo. In this accident, all 133 people aboard the plane (both passengers and crew) were killed. Immediately after the accident, an Accident Investigation Committee was set up by the Ministry of Transportation. The committee submitted reports, including the results of a simulation which was voluntarily undertaken by one of the committee members [9]. For us, there is no way of knowing the exact cause of the accident. However, the official reports suggest that the most probable cause of the accident was complete failure of the engine due to fatigue failure of the engine set bolts (called 'cone bolts'). The failure must have caused the plane to lose its balance and crash into the sea.

In July 1975, the hook of a crane in use at a certain steelworks broke, resulting in the death of employees. The accident is attributable to the fatigue failure of the screws. In any case, the screws are the most critical parts in the crane hook. For this type of failure, the report prepared by Kitsunai [10–12] will be of good reference.

Apart from such serious accidents, failures of bolts occur frequently [1,10–16]. In Figure 1.4 (Section 1.3), bolt failures stand third in importance but the number of bolt failures that actually occurred is estimated to be the largest. Bolts can be more easily manufactured and purchased than other parts. If they are broken, therefore, they are simply replaced by new bolts. For this reason, the number of failures reported is far smaller than the number of failures which have actually occurred.

In modern plants, the operations are performed on-line to increase production efficiency. It may not be too much to say that the smooth operation of machines at these plants is dependent largely upon a single bolt. Since failure of a bolt is attributable mostly to fatigue, designing against fatigue is very important for bolts.

In this section, several examples of bolt failure are described after a general explanation of the fatigue strength of bolts. Finally, the importance of bolts is emphasized by explaining a method for improvement of the fatigue strength of bolts.

4.7.1 Fatigue limit of ordinary bolts

Only a little work has been directed towards bolt fatigue. The tensile fatigue limits [17] of representative steel bolts are shown in Table 4.16. According to the definition of fatigue limit, the fatigue limit means a limiting stress below which an infinite number of stress cycles can be applied. In the case of partially tensile

Table 4.16 Tensile fatigue limit of steel bolt

Nominal diameter, d(mm)	6	8	12	20	30	42	48
Fatigue limit, σ_w (kgf/mm²)	6	8	5	4	3	3	3

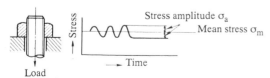

Note: More strictly, the fatigue limit is indicated in terms of $\sigma_a + \sigma_m$ by definition.

Table 4.17 Tensile fatigue limit of bolt

King of steel	σ_{wB} (kgf/mm2)	σ_{w0} (kgf/mm2)	σ_B (kgf/mm2)	β	α
0.3% C	40.3	26.0 (13.0)	14.8 (7.4)	1.76	3.86
SAE2320	76.6	51.3 (25.7)	15.5 (7.8)	3.32	3.86

σ_B Tensile strength
σ_{w0} Fatigue limit of plain specimen
σ_{wB} Fatigue limit of bolt
β Fatigue notch factor (fatigue limit of plain specimen/fatigue limit of notched specimen)
α Stress concentration factors (by photoelastic method)

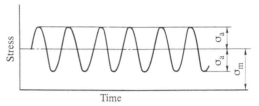

Pulsating tensile stress

Note: As the fatigue limit is the maximum stress which stands infinite repetition, it becomes the sum $\sigma_a + \sigma_m$ of the stress amplitude and mean stress in the case of pulsating tensile stress. When discussing the fatigue properties of a bolt, it will be useful to compare the stress amplitude σ_a values shown in ( ), because the effect of mean stress is small.

pulsating fatigue, the fatigue limit should be expressed in terms of the sum of stress amplitude σ_a and mean stress σ_m. In the case of bolts, however, the effect of mean stress is comparatively small, as described later (see Figure 4.65) and the mean stress is not always constant. Accordingly, a comparison in terms of stress amplitude σ_a only will facilitate understanding. In this book, therefore, the fatigue limit is expressed in terms of stress amplitude only unless otherwise specified. From Table 4.16 the tensile fatigue limit of steel bolts of normal diameters is 5–6 kgf/mm² but decreases with increasing nominal diameter. This decrease is called the size effect, which is particularly significant in the case of fatigue failure. However, as the size effect for fatigue failure of steel structures is only 10–15% in terms of decrease in fatigue limit [18], it is considered that the decrease in fatigue limit of a bolt is very large. The probable reason is localized loading between the bolt threads and the nut threads due to low machining accuracy, which is one of the factors governing the fatigue strength of bolts as described in Section 4.7.3 (S. Nishida and C. Urashima, 1982, unpublished data).

Table 4.17 shows the tensile fatigue limits of bolts excerpted from other literature [19]. From Table 4.17 the fatigue limit of a plain specimen increases with increasing tensile strength but the fatigue limit of a bolt hardly varies (see Figure 4.67). This may be because

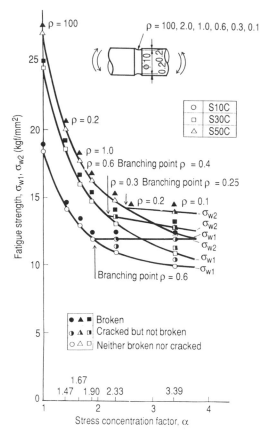

Figure 4.65 Relation between stress concentration factor α and fatigue strength σ_{w1} and σ_{w2}

the bolt is a kind of notched specimen. The fatigue limit of a plain specimen increases with increasing tensile strength. In the case of a notched specimen, the notch sensitivity increases with increasing tensile strength, resulting in a sharp decrease in fatigue limit. Accordingly, the difference in tensile strength does not have a noticeable effect on the fatigue limit of a notched specimen (see Figure 4.65). Moreover, the above phenomenon is also attributable to the same cause as that of the size effect of a bolt (see Figure 4.67).

The effect of mean stress on the tensile fatigue limit of bolts is shown in Figure 4.66 [20]. So far as Figure 4.66 is concerned, the mean stress below about 40 kgf/mm² has little effect on the fatigue limit. However, some work reports that the average stress has little effect on bolts with low tensile strength but shows its effect on bolts whose tensile strength has been increased by heat treatment [21]. According to an experiment conducted by the authors, the fatigue limit of bolts with a tensile strength $\sigma_B = 110$ kgf/mm² (quenched and tempered structure) is decreased by about 20% when the mean stress σ_m is increased from 18 to 56 kgf/mm² (S. Nishida and C. Urashima, 1982, unpublished data).

Figure 4.67 shows the relation between the tensile strength and the tensile fatigue limit of bolts. The tensile fatigue limit varies considerably but the tensile fatigue limit increases, albeit only slightly, with increasing tensile strength. These variations in the fatigue limit of bolts are attributable to the transmission of force due to contact between the bolt threads and the nut threads.

Figure 4.68 shows the size effect on the tensile fatigue limit of bolts [20]. The tensile fatigue limit

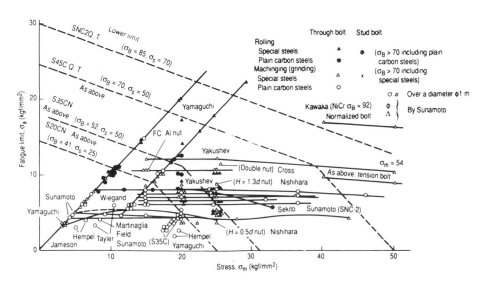

Figure 4.66 Effect of mean stress on fatigue limit of bolts

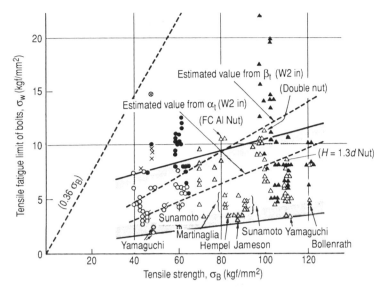

Figure 4.67 Relation between tensile strength and tensile fatigue limit (see Figure 4.66 for key)

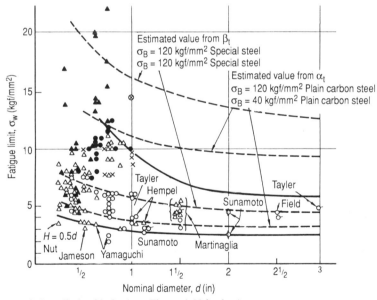

Figure 4.68 Size effect on fatigue limit of bolts (see Figure 4.66 for key)

decreases with increasing nominal bolt diameter, but the tensile fatigue limit varies considerably (see Figures 4.66 and 4.67). From Figure 4.68 the tensile fatigue limit of a bolt 3 inches in nominal diameter will be only 2.5–3.0 kgf/mm² if a conservative value is desired. This value is very low compared with the tensile strength. The shape factor (stress concentration factor) α of a bolt is about 4 (see Table 4.17). The notch factor β is expressed as 'the fatigue limit of a plain specimen, σ_{w0}/fatigue limit of a notched specimen, σ_w.' Normally, the stress concentration factor α of a

notched specimen is nearly equal to β when this factor is small. If α is greater than 2, α becomes greater than β. In the case of bolts, however, the notch factor β becomes 8–10 and β becomes greater than α even if the fatigue limit of a plain specimen with a tensile stress $\sigma_B = 100$ kgf/mm² is assumed to be $\sigma_{w0} \fallingdotseq \sigma_B/2 = 50$ kgf/mm², because the fatigue limit, σ_w, of a bolt is 5–6 kgf/mm². This indicates that the fatigue limit of bolts is far lower than that of conventional notched specimens. Conversely, improvement of the fatigue limit is far more difficult for bolts than for notched specimens.

Table 4.18   *Chemical composition in the vicinity of position of failure* (wt %)

No.	Location	C	Si	Mn	P	S	Cu	Ni	Cr	Mo	Al
3	Centre	0.44	0.36	0.78	0.015	0.033	0.15	0.08	1.06	0.23	0.003
2	Intermediate	0.41	0.36	0.75	0.014	0.031	0.15	0.08	1.05	0.22	0.008
1	Surface (crack initiation)	0.41	0.35	0.73	0.013	0.020*	0.14	0.08	1.02	0.21	0.006
	Specification	0.37–0.44	0.15–0.35	0.55–0.90	<0.030	<0.030	<0.030	–	0.85–1.25	0.15–0.35	–

* Large scattering 0.018–0.030.

4.7.2 *Failure of fastening screws*

(a) **Failure of the tie rod of a rolling mill**
The front view of a rolling mill and the position of failure in the mill are shown in Figure 4.69. The mill is designed so that the rolling reaction force is received by four giant bolts (called the tie rod – outside diameter of thread = 478 mm, $\phi\,450 \times 13\,975$ mm total length). Failure was detected in two out of four tie rods. One of the two tie rods was completely broken and cracks were detected in the other rod. This failure is shown schematically in Figure 4.70. In the two tie rods, failure occurred in the end section of the nut or its vicinity on the side where the material to be rolled is bitten. The material of the tie rod is SCM440. The chemical composition of SCM440 is shown in Table 4.18.

The fracture surface is shown in Figure 4.71. For the most part the fracture surface is brittle, while a smooth surface with a step is observed near the thread root. This surface is attributable to the fatigue crack which initiated at the thread root and propagated. Observation by SEM revealed striation peculiar to fatigue

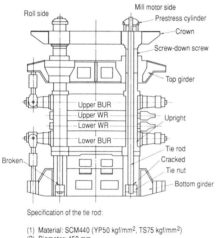

Specification of the tie rod:

(1) Material: SCM440 (YP50 kgf/mm², TS75 kgf/mm²)
(2) Diameter: 450 mm
　　Length: 13.975 mm
　　Weight: 17.7 tf
(3) Prestress: 2685 tf/bolt
　　Oil pressure: 800 kgf/cm² → 16.9 kgf/mm²
(4) Yielding load: 8200 tf/bolt

Figure 4.69   Schematic illustration of rolling mill and position of failure

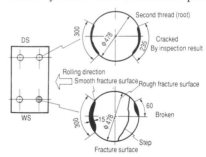

Figure 4.70   Schematic illustration of failure

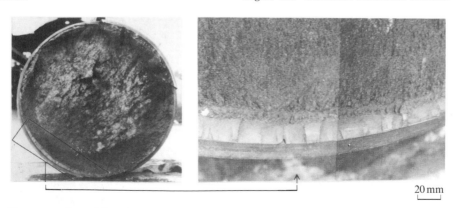

Figure 4.71   Fracture surface of tie rod   Enlarged view

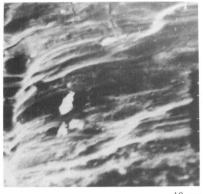

Figure 4.72 Results of observation of fracture surface by SEM

(Figure 4.72). After repair, the stresses were measured by strain gauges (single axis × 5 mm long, four gauges) which were attached to the body of the tic rod. As a result, a maximum repeated stress amplitude $\sigma_{amax}=1.23$ kgf/mm^2 and a mean stress $\sigma_m=17.7$ kgf/mm^2 were obtained. When the coefficient of internal force ϕ is assumed to be about 0.2 [48], the amplitude of external force, F_{amax} is estimated to be about 6.2 kgf/mm^2. As the tensile strength σ_B of the material is 80.7 kgf/mm^2, the fatigue limit is as low as about 1/50–1/60 of the actual tensile strength. The reason why the fatigue strength of large-diameter bolts decreases sharply is explained in Section 4.7.3.

(b) Failure of a cross-head pin holding the bolts of a compressor

The cross-head pin holding the bolts of the connecting rod of a double-action compressor broke. Five out of six bolts in use were broken. A bolt is shown schematically in Figure 4.73. The piston rod was used for 52 days at a frequency of about 3000 cycles/min. According to the reply from the manufacturer to our inquiry which was made before the investigation, the bolts are made of S25C. As a result of the failure, the connecting rod was bent and the casing was broken. Since a serious accident was caused by the failure of a single bolt, a detailed investigation was made.

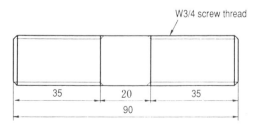

Figure 4.73 Dimensions of cross-head pin holding bolt of compressor

The items investigated were as follows:

1. Chemical analysis
2. Macroscopic observation and sulphur print
3. Microscopic structure
4. Hardness distribution
5. Macroscopic and microscopic observation of the fracture surface
6. A dynamic study
7. A synthetic study.

The results of the investigation are described below (although part of the results is omitted.)

Figure 4.74 shows the sampling procedure of the broken bolt. The stud bolt was broken at the end section of the external thread.

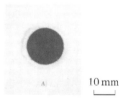

Figure 4.74 Sampling procedure for specimens

Figure 4.75 shows a sulphur print of the cross section of the broken bolt. (The sulphur print method investigates the macroscopic distribution of sulphur in the cross section by transferring it to the printing paper moistened with dilute sulphuric acid. Refer to JIS G0560.) As the specimen used was thin, it was embedded in resin to facilitate handling. The resin is seen as an annular ring in the periphery. From Figure 4.7 the sulphur content is seen to be very high.

Figure 4.75 S-print (transverse section of specimen shown in Figure 4.74)

The chemical composition of the bolt is shown in Table 4.19. According to the results of chemical analysis, the bolt is made of sulphur free-cutting steel. The analysis indicated that the material of the bolt is not S25C as insisted by the manufacturer but a steel equivalent to SUM23. This is a good example to show that the manufacturer's explanation is not always correct. Special attention should be paid to this point.

Table 4.19   *Chemical composition of bolt* (wt %)

C	Si	Mn	P	S	Cu	Ni	Cr	Al
0.082	0.010	1.05	0.060	0.305	0.011	0.030	0.049	Trace

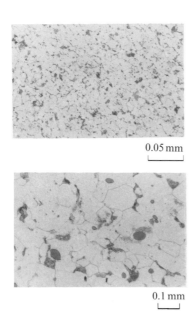

Figure 4.76   Optical microstructure (transverse section)

An optical micrograph is shown in Figure 4.76. The black rectangular points are MnS. MnS is observed all over the surface. The maximum size of MnS is about 20 μm. The structure is mostly ferritic with a small quantity of pearlite. The carbon content of the steel is low.

Figure 4.77 shows the hardness distribution in the cross section of the bolt. The mean hardness H_v is 180–200. The fracture surface is shown in Figure 4.78. In

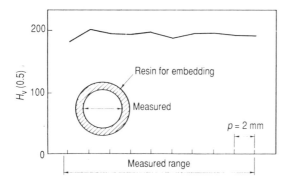

Figure 4.77   Vickers hardness distribution (transverse section)

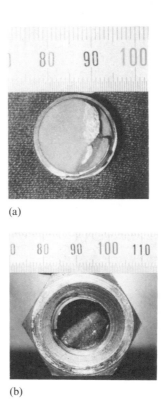

(a)

(b)

Figure 4.78   Fracture surface: (a) bolt A; (b) bolt B

the bolt A, the main point of fracture initiation is on the left, and the crack propagated from the left to the right side. No beach marks are observed on the fracture surface. The bolt B is an example in which fracture occurred at the end section of the nut. The ratio of fatigue failure of the bolt is small. It is estimated that the bolt was broken as the load applied to it was considerably increased because of the earlier failure of several other bolts. The failure initiated at more than two points.

Examples of the results of observation of the fracture surface by SEM are shown in Figures 4.79 and 4.80. Figure 4.79 shows the crack initiation point. The small black points are MnS. Figure 4.80 shows the striation observed in part of the fracture surface. The macroscopic direction of crack propagation is from left to right.

The stress applied to the bolt is estimated on the basis of fracture mechanics as described below. Figure

Direction of crack propagation ⟶   0.5 mm

Figure 4.79 Results of observation of fracture surface by SEM

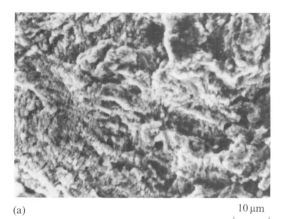

(a)   10 μm

Direction of crack propagation

⟶

(b)   5 μm

Figure 4.80 Results of observation of fracture surface by SEM: (a) position 8.7 mm from the thread root; (b) enlarged view of (a)

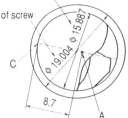

Figure 4.81 Schematic illustration of fracture surface and point A observed by SEM

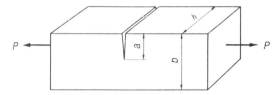

Figure 4.82 Model of bolt with fatigue crack

4.81 shows schematically the fracture surface and the position of observation by SEM. A model for the estimation of stress is shown in Figure 4.82.

From Figure 4.80 the striation distance S at point A is 6.7×10^{-4} mm/cycles. Accordingly, K_I is calculated as nearly equal to 110 kgf/mm$^{3/2}$ on the basis of the relation between the striation distance and the crack propagation rate da/dN [2–4] and the relation between the crack propagation rate da/dN and the stress intensity factor K_I [5,6].

From the crack propagating condition in the fracture surface, it is apparent that the crack propagates in one direction only from one side (point C in Figure 4.81) to point A. Accordingly, a calculation is made from the model shown in Figure 4.82.

Outside diameter of bolt $d_o = 19.004$ mm
Root diameter of thread $d_i = 15.887$ mm
$h = d_i$
$b = (d_o + d_i)/2 = 17.446$ mm   (4.28)
$a = (d_o - d_i)\beta + 8.7 = 10.259$ mm $(\beta' = 0.5)$ (4.29)

a_o is the initial crack length. The K values for bolts with a notch as shown above are not available. For convenience, half of the thread height is used as the K value. A study is now under way on this point. At present, the factor β of the first term of equation (4.29) is taken as 0.5. Exact calculation of this factor will be indispensable in the future. Accordingly, the stress intensity factor K_I is given by the following equation [23]:

$$K_I = \frac{P\sqrt{a}}{bh\sqrt{\pi}}\left[1.99 - 0.41\left(\frac{a}{b}\right) + 18.70\left(\frac{a}{b}\right)^2\right.$$
$$\left. - 38.48\left(\frac{a}{b}\right)^3 + 53.84\left(\frac{a}{b}\right)^4\right] \tag{4.30}$$

The nominal stress σ_0 in the minimum cross section is

$$\sigma_0 = P/(b-a_0)h \qquad (4.31)$$

Substituting the values shown above into equation (4.31)

$$\sigma_0 = 9.926 \text{ kgf/mm}^2 \qquad (4.32)$$

Assuming that the stress amplitude σ_a is half the value shown above, σ_a is estimated as shown below:

$$\sigma_0 = 5.0 \text{ kgf/mm}^2 \qquad (4.33)$$

According to the results of a laboratory test which was conducted separately, the fatigue limit σ_w of the SUM23 bolt is 5.5 kgf/mm², which is close to the value shown above.

The probable reasons why the fatigue limit determined experimentally is slightly greater than the limit calculated from the fracture surface are as follows. One reason is that for convenience a bolt 220 mm in overall length (double end) was used in the laboratory test. The use of a long bolt is advantageous for the estimation of fatigue characteristics in cases where a bending force is applied to the bolt. Another reason is the difference in the threading condition. Moreover, errors in the localized measurement of striation distance must be taken into account. Furthermore, a stress very close to the fatigue limit of the bolt was applied to the bolt during service.

As described in Section 4.7.2(a), the external force can also be calculated from the striation distance. Calculation by this method is recommended to the reader.

(c) Failure of a cylinder rod

The cylinder rod of a hydraulic upsetter broke from the threads. The cylinder rod is shown in Figure 4.83. The macroscopic fracture surface of the rod is shown in Figure 4.84. The fracture occurred at the end of the engagement with internal threads. The material of the rod is S50C (quenched and tempered steel). As is

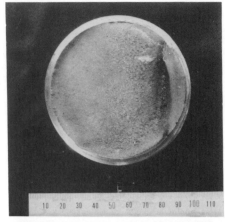

Figure 4.84   Cylinder rod of upsetter

apparent from the macroscopic fracture surface, the ratio of the fatigue fracture surface is very high. It is therefore estimated that the repeated stress amplitude σ_a is close to the fatigue limit of the bolt. The number of cycles to failure actually measured, N_f, is 25×10^4. With QT steels, striation is not always clearly observed on the fracture surface under SEM. Accordingly, an example of analysis from the fatigue strength (S–N diagram) is described below.

First, the fatigue limit of the screw now in use is estimated. In the case of screw M90, the root diameter of the thread is 83.5 mm. As a load P of 0–57.4 tf (a constant amplitude of load generated by hydraulic pressure) is repeatedly applied, the nominal stress amplitude σ_a is calculated as shown below:

$$\sigma_a = \frac{57.4 \times 10^3}{2A} = 5.24 \text{ kgf/mm}^2 \qquad (4.34)$$

where A is the root area.

Taking the size effect into account, it is estimated that the fatigue limit of the screw is lower by about 20% than that of a standard bolt M24. If the value calculated by equation (4.34) is converted into the

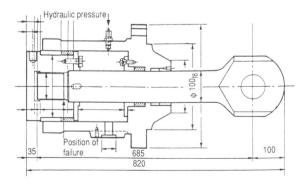

Figure 4.83   Schematic representation of cylinder rod (QT steel of S50C)

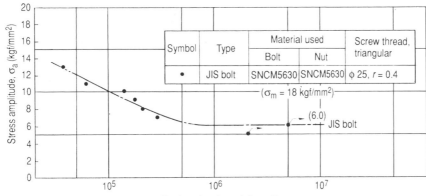

Figure 4.85 *S–N* curve for normal bolt

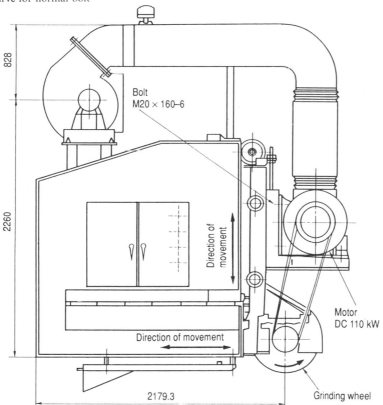

Figure 4.86 Schematic illustration of grinder (side view)

value for an M24 bolt, the stress amplitude σ_{a1} is as shown below.

$$\sigma_{a1} = 5.24/0.80 = 6.55 \text{ kgf/mm}^2 \qquad (4.35)$$

The *S–N* curve of a normal bolt is shown in Figure 4.85. When the stress level is as calculated by equation (4.35), the number of cycles to failure, N, is 40×10^4 cycles. The order of this number is the same as that of the number of cycles to failure actually measured, N_f (25×10^4).

(a) Failure of grinder set bolts

There are cases in which repeated stress is not apparently applied to the bolt or in which it is difficult to predict the stresses to be applied to the bolt at the design stage. An example of bolt failure in such cases is described below.

A grinder is shown schematically in Figure 4.86. Bolts are used in many places in the grinder. The bolt most likely to be broken is the bolt (M20 × 160 mm in length, six bolts) fixing the motor. A grinding wheel is

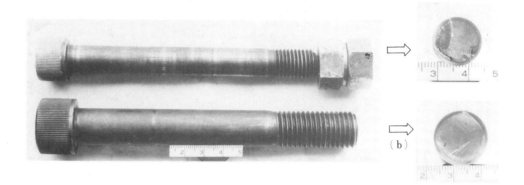

Fracture surface

Figure 4.87   Example of broken fixing bolts for grinder

(a)

(b)

Figure 4.88   Outer view of compressor piston rod: (a) compressor being disassembled for repair; (b) unbroken compressor piston rod

located beneath the motor (DC 110 kW). The motor is connected to the grinding wheel with four V-belts. The motor and the grinding wheel are fixed on a common slider which is designed to move up and down. The billet is sent under the grinding wheel as if it crosses the wheel, and the billet surface is partially ground by the wheel.

Apparently, little repeated load is applied to the motor fixing bolts. In practice, however, the impact force produced when the grinding wheel comes into contact with the billet is transmitted through the V-belts to the motor, as a result of which fluctuating loads are applied to the bolts. Moreover, the vibrations resulting from the rotation of the motor, V-belts, and grinding wheel are applied to the bolts in addition to the fluctuating loads.

An example of failure of a grinder set bolt is shown in Figure 4.87. Almost the entire surface is occupied by the fatigue fracture. Judging from the above description and the high ratio of the fatigue fracture surface, it is estimated that the level of repeated stress was close to the fatigue limit of the bolt.

When bolt failure started to increase, the set bolt was replaced by a new bolt with higher tensile strength σ_B of 80 kgf/mm^2, but the fatigue life of the new bolt was nearly half that of the old bolt. Accordingly, the material of the bolt was changed to soft steel (SS41) as a drastic measure. As a result, the fatigue life of the bolt

was prolonged by about 50% compared with that of the old bolt.

As described above, even bolts for which the generation of only little repeated stress is expected at the design stage may be broken due to fatigue. Depending on the individual case, bolts with low strength may have a longer life. As these points are contrary to common sense, they will be explained in more detail in Section 4.7.4.

(e) Failure of a compressor piston rod

If one thread of a bolt is broken, the breakage may spread to other parts in succession. A good example is the failure of a compressor piston rod. The compressor is of the double acting type (about 2000 kW). As the screw (M90) of the piston rod on one side was broken, the crank shaft lost its balance, resulting in failure of the balance weight shear pin, bearing clamp bolt, connection rod, and compressor casing bed in the order mentioned.

The compressor piston rod (reciprocating type) is shown in Figure 4.88. Shown here is the piston rod on the side which was not broken. The colour check of the rod did not reveal fatigue cracks. The fracture surface of the piston rod is shown in Figure 4.89. Figure 4.89(c) is the longitudinal cross section of the rod shown in Figure 4.89(b). The failure of the piston rod occurred from the end section of the nut. As the piston

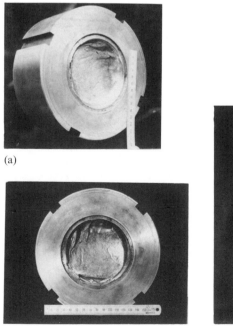

(a)

(b)

(c)

Figure 4.89 Outer view of fracture surface of compressor piston rod: (a) broken piston rod; (b) fracture surface of (a); (c) vertical section of (b)

rod is of the reciprocating type, its fracture surfaces were severely struck by each other and could not be used for observation. Quite a long crack was observed at the root of the twelfth thread, counting from the end section of engagement with the nut (Figure 4.89(c)). It may be said that the design of the threads of the piston rod is quite reasonable.

The material of the piston rod is SNCM625 (quenched and tempered steel, tensile strength $\sigma_B = 94.4\ \mathrm{kgf/mm^2}$). The microstructure is a tempered martensitic structure. When the rod was broken, about five months were required for complete repair.

(f) Failure of the chisel holding bolt of a large machine drill

An example of the failure of bolts for machines used in the civil engineering and construction fields is described below.

Figure 4.91 shows a large machine drill in which the

Figure 4.90 Outer view of chisel holding bolts of large stone crusher

chisel holding bolt was broken. Figure 4.90(b) shows the setting condition of the bolt (M50 × length 1200 mm). Figure 4.91 shows the appearance and fracture surface of the bolt. Four bolts make one set. It is considered that the reaction force of the inertia force produced by the reciprocating motion of the chisel and the impact force transmitted through the chisel are applied to the bolts. The bolts were broken during a period of six months to a year, although the duration varies depending on the method of use.

(a)

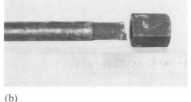

(b)

(c)

Figure 4.91 Chisel holding bolt of large stone crusher: (a) appearance of bolt; (b) broken; (c) fracture surface

The beach mark which is characteristic of fatigue is clearly seen in the fracture surface shown in Figure 4.91. The ratio of the fatigue fracture surface is higher than 80% and the fracture surface is smooth and nearly free from steps. It is therefore considered that the repeated stress is close to the fatigue limit.

(g) Failure of the set bolt of a rod mill used for crushing

The rod mill consists of a horizontally installed cylinder in which a rod of length nearly equal to the mill length is inserted. As the cylinder is rotated around its axis, the rod is raised and dropped to crush the stone material. A steel plate with excellent wear resistance is attached with bolts to the inside wall of the rod mill. The bolt which was broken is the liner set bolt (M40 × length 120 mm) of the rod mill which is used for crushing materials for cement. The impact resulting

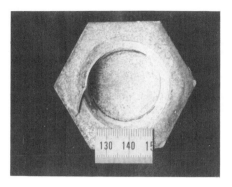

Figure 4.92 Fracture surface of fitting bolt for crushing rod mill

from dropping the rod is transmitted to the mill liner, causing failure of the bolts fixing the mill liner.

The fracture surface of the liner set bolt is shown in Figure 4.92. The fracture surface is a typical fatigue fracture surface and the ratio of the fatigue fracture surface is nearly 100%. It is therefore considered that fatigue cracks initiated and propagated at some distance and that the external force was decreased because of the shifting of part of the load to other bolts. In such cases, fatigue fracture spreads from one bolt to another in succession.

The bolt is made of SCM435 (quenched and tempered steel, tensile strength $\sigma_B = 100 \, \text{kgf/mm}^2$). Failure occurred at the end section of the nut. As lime powder is adhering to the bolt, the whole bolt looks whitish.

(h) Failure of the set bolt of a universal fatigue tester

There is a proverb that the shoemaker's wife goes barefoot. The universal fatigue tester is used to evaluate the fatigue characteristics of test specimens. Accordingly, the tester receives the reaction force of the load applied repeatedly to the specimen. Although test specimens are changed one after another, the tester itself is not changed. When the tester is used for a long

Figure 4.93 Hydraulic actuator of universal fatigue tester

period of time, therefore, failure of the tester parts occurs frequently.

Figure 4.93 shows the hydraulic actuator of a universal fatigue tester. In the actuator, eight bolts each are used for the top and bottom covers (double end, W1-1/4). As oil leakage from the bottom cover was detected, the bolts for the bottom cover were checked. As a result, failure of three out of eight bolts was detected. The fracture surface of the set bolt of the actuator holding cover is shown in Figure 4.94. A

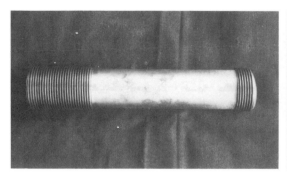

10 mm

Figure 4.94 Fracture surface of set bolt of hydraulic actuator holding cover of universal fatigue tester (thread: W1.1/4 inch)

fatigue fracture surface is clearly seen all over the surface. The failure occurred from the end of the engagement with internal threads. The material of the bolt is SCM435 (quenched and tempered steel).

Figure 4.95 shows the fracture surface of an anchor bolt of a universal fatigue tester which is of the same type as that shown in Figure 4.93. The failure of the bolt occurred from the end section of the nut. The material of the bolt is SS41. Failure occurred in one of four bolts (M24).

Figure 4.95   Fracture surface of anchor bolt of fatigue testing machine

This type of fatigue tester is so constructed that the external force and reaction force are received within the frame. So far as desk calculation is concerned, the force is not applied to the anchor bolts at all. As the tester had been in use for 12 years, the failure of the anchor bolt may be attributed to the vibration of the tester proper, which was caused as the hydraulic actuator becomes loose. As described above, the deterioration of machines with age is one of the causes of bolt failure.

(i) Failure of a jig (column) for the reversed bending test

As described in (h) above, fatigue failure of the members may occur even if their fatigue life is closely examined at the design stage. A jig for the reversed bending test and an example of failure of the jig are shown in Figures 4.96 and 4.97 respectively. The bolt is M85 × 750 mm in length (root diameter: 78.5 mm). Failure occurred from the rod at the end section of the internal threads. The material of the jig is SCM435 (quenched and tempered steel).

As the reaction force in three-point bending is applied to the bolt (column), it is considered that not only a tensile stress but also a bending stress were applied to the threads. This is evident from the fracture surface shown in Figure 4.97(d).

The failure occurred on the triangular thread side.

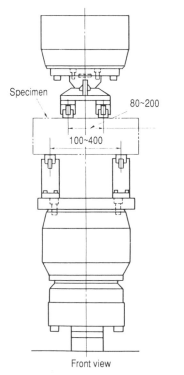

Front view

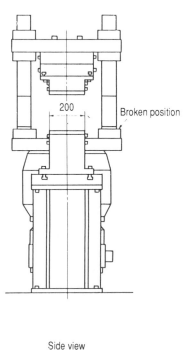

Side view

Figure 4.96   Equipment for repeated bending test: (a) front view; (b) side view

The reason why failure did not occur on the square thread side (right-hand side of Figure 4.97(a), root diameter: 72.4 mm) in spite of the smaller root diameter is that the position of engagement with the nut on the square thread side is changed at regular intervals for adjustment. When a repeated load is applied, one of the simplest methods for improving the fatigue strength is to change the position of engagement with internal threads at regular intervals, although this method is not widely employed. A beach mark was not observed on the fracture surface of this bolt. This shows that the test was conducted with a nearly constant stress amplitude.

(j) Failure of a track bolt for the fish plate of a rail

At rail joints, a clearance of several millimetres is provided for rail expansion and contraction due to differences in temperature.

Fish plates are used to reinforce the rail joints. The fish plate is clamped with four to six track bolts (M24 × about 180 mm in length). When a wheel rolls on the rails, an impact force is applied to the rails at the joints. Accordingly, tensile and shear stresses are created in the bolt.

An example of failure of a track bolt (material: SS41) is shown in Figure 4.98. The failure initiated at incomplete threads and is a fatigue failure.

(a)

(b)

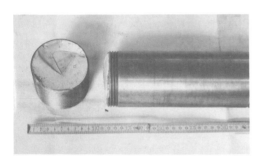

(c)

Figure 4.98   Track bolt for fishplate of rail (failure from incomplete threads)

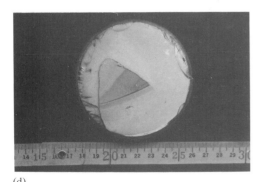

(d)

Figure 4.97   Broken portion of repeated bending equipment (column): (a) outer appearance; (b) position of break; (c) fracture surface; (d) enlarged view of (c)

Generally, the failure of bolts occurs at three points, i.e. the end of engagement with internal threads, incomplete threads, and the underhead fillet. The probabilities that the bolt is broken from these points are said to be 65, 20 and 15%, respectively [10,11]. However, so far as the fatigue failure of clamping bolts is concerned, more than 80–90% of failures occur from the end of engagement with internal threads unless special means for improvement of fatigue strength are taken. The probable reason why the failure of this track bolt occurred from the incomplete threads is that not only a uniaxial tensile force but also a shear force was applied to the bolt.

4.7.3 Summary of failures of fastening screws

The fastening screw which is represented by the combination of bolt and nut is an important part of machines and equipment. However, it seems that this importance is not fully realized, but they are regarded as one of simple consumable parts. In practice, however, the failure of even a single bolt may have a great effect depending on the purpose of its use. In the preceding sections, several representative examples of bolt failure have been described. The importance of function and safety design of fastening screws should be emphasized once again. The foregoing description may be summarized as follows.

1. Let α be the stress intensity factor and β the notch factor. In the fatigue failure of plain specimen, α is always equal to or greater than β. In the case of a bolt, it is not rare for α to be about 4 but β is 8–10. There are many other cases in which α becomes smaller than β. For calculation of the fatigue limit of bolts, therefore, it is very dangerous to make a calculation on the basis of α only as is being done for notched specimens.
2. In general, the fatigue limit of bolts is 5–6 kgf/mm². The limit decreases with increasing nominal diameter. The so-called size effect on the fatigue limit of a bolt is far greater than that on a notched specimen. The fatigue limit of a bolt is little influenced by the mean stress (clamping force) and static tensile strength. These values are likely to vary widely. At the design stage, therefore, it is desirable to adopt the results of the fatigue test of a bolt.
3. Even in cases where little repeated load is apparently applied to the bolts, the bolts may break because they vibrate as a result of their looseness or age deterioration of machine parts connected to a source of vibration.
4. It is important to prevent the looseness of fastening screws. It is quite effective for the extension of fatigue life to use fastening screws made of a comparatively soft material (internal threads in particular) or to shift the position of engagement with internal threads at regular intervals.

The reasons why these items were taken up in this book are described in Section 4.7.5.

4.7.4 Measures taken so far for the improvement of fatigue strength of fastening screws and their effects

(a) Causes of low fatigue strength

There are several causes of low fatigue strength of bolts. The first cause is uneven load sharing [24] among the threads of a bolt (see Table 4.20 and Figure 4.99). In the case of a bolt with eight threads for engagement with the nut, about one-third of the total load (which is taken as 100%) is applied to the first thread, as is seen from Table 4.20. The loads applied to the second and ensuing threads decrease sharply. The ratio of the load applied to the fourth and ensuing threads in particular is less than 10%. The same also applies to bolts with six or ten threads. It is estimated that this type of load is applied to nearly half the height of each thread in a concentrated manner.

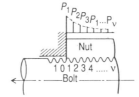

Figure 4.99   Load distributed to screw threads

On the basis of the load-sharing condition shown in Figure 4.99, the stress developed at each thread root is explained by reference to Figure 4.100. The loads to be shared by the threads are represented by $P_1, P_2, P_3 \ldots P_v$, beginning with the first thread in engagement with the nut thread. The stresses developed at the roots of the first thread, second thread ... and vth thread by the load P_1 are represented by $+\sigma_{11}, -\sigma_{12} \ldots -\sigma_1 v$, respectively. Similarly, the stresses developed at the roots of the first, second ... vth threads by the load P_2 are represented by $+\sigma_{21}, +\sigma_{22}, -\sigma_{23}, \ldots \sigma_2 v$, respectively. In general, the stress developed at the root of the nth thread by the load P_m is σ_{mn}. Here, '+' indicates the tensile stress, while '−' indicates the compressive stress. Accordingly, the stress induced when $m \geq n$ is a tensile stress, but a compressive stress is created when

Table 4.20   *Percentage load distribution to screw threads*

Number of thread	P_1	P_2	P_3	P_4	P_5	P_6	P_7	P_8	P_9	P_{10}
6	33.7	22.9	15.8	11.4	8.7	7.5				
8	33.3	22.3	15.0	10.2	7.0	5.0	3.9	3.3		
10	33.1	22.2	14.9	10.0	6.7	4.6	3.1	2.3	1.6	1.5

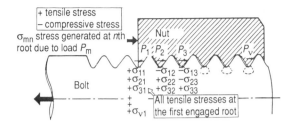

Figure 4.100 Stresses generated at the roots of bolt threads

$m < n$. The stress developed in a single bolt is the sum of the stresses developed at the roots of all the threads. The stresses induced at the roots of the second and ensuing threads under the load-sharing condition as shown in Figure 4.69 are decreased as the compressive and tensile stresses cancel each other to some extent. However, only a tensile stress is induced at the root of the first thread at the end section of the nut. It is therefore apparent that the stress developed at this root is the largest. The above description is substantiated by the results of experiments conducted by Seike, Sakai and Hosono [25] on the stress concentration in bolts (see Figure 4.101). In other words, the stress concentration factor α at the root of the first bolt thread at the end section of the nut is 4.5, which is close to the value ($\alpha = 3.86$) obtained by the photoelastic test, as shown in Table 4.16. It is therefore easily understood that bolt failure mostly occurs at the root of the first bolt thread at the end of engagement with the nut. In general, bolt failure occurs at three points, i.e. the end face of the nut, incomplete threads, and the underhead fillet. The probabilities that failure occurs at these points are said to be 65, 25 and 10 [19]. So far as the fatigue failure of clamp bolts is concerned, more than 80–90% of failures occur at the root of the first bolt thread at the end section of the nut unless special measures are taken for the improvement of fatigue

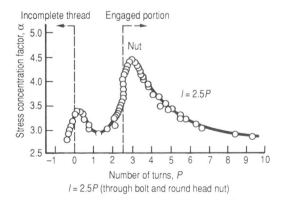

Figure 4.101 Mutual interference of stress concentration at root of bolt

strength. As is apparent from the examples of failure of fastening screws described in Section 4.7.2, all failures, except one example (Figure 4.97), occurred at the root of the first bolt thread at the end section of the nut.

Uneven load sharing has been cited as a cause of low fatigue strength of bolts. As the bolt is a kind of notched material and the external force is transmitted through contact between the bolt threads and the nut threads, a high stress concentration factor and localized loading can also be cited as causes of low fatigue strength of bolts. These causes are explained in more detail in Section 4.7.6.

The measures taken so far for improvement of the fatigue strength of fastening screws are roughly divided into measures for nuts and measures for bolts. The measures are described below.

(b) Measures for nuts

Figure 4.102 shows the transmission of force from the bolt to the nut [19]. In general, complicated stress concentration can be quantitatively determined by likening it to the flow of water. Figure 4.102(a) shows a conventional type of fastening with a bolt and nut. From Table 4.19 and Figures 4.99–4.102(a) it is seen that the stress is concentrated at the boundary between the bolt and the nut. To improve the fatigue strength of

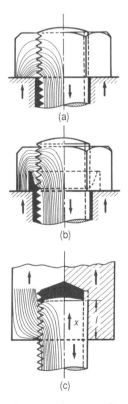

Figure 4.102 Stress flow from bolt to nut

a bolt, therefore, the stress concentration at the nut end should be decreased. One such measure is to change the flow of stress by providing a weir at the nut side, an example of which is shown in Figure 4.102(b). That is, the flow of stress to the end of the nut (the end of the bolt threads) is changed by providing an annular groove at the nut end. The most desirable method is to pull the nut as shown in Figure 4.102(c). The smooth flow of force from the bolt to the nut can be ensured by pulling the nut. The bolt is elongated but the nut is compressed under a tensile force. As the 'strain' is concentrated at the nut end due to the difference in deformation between the bolt and nut, the fatigue strength of the bolt is decreased. If the strain concentration is decreased, the fatigue strength of the bolt will be considerably improved. It seems that the idea of 'nut pulling' is based on this idea.

Figure 4.103 shows examples of the measures proposed for nuts so far [19,26]. Strictly speaking, the example shown in Figure 4.103(e) is not a measure for nuts. However, the measure can be included in the measures for nuts as it is based on the same idea as other measures. It will be easily understood that the measures shown in Figure 4.103(a), (b), (c) and (d) are based on the measures shown in Figure 4.103(b) or (c). Although the measure shown in Figure 4.103(e) is a measure for bolts, the measure aims at ensuring equal elongation of the bolt at the point where it engages with the nut.

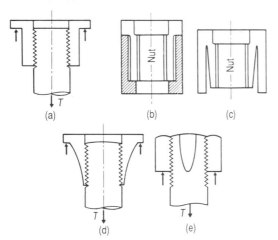

Figure 4.103 Conventional methods for improving the fatigue strength of nuts

However, the measures described above have not been put into practice for the following reasons. With the measure shown in Figure 4.103(a), for example, a fatigue test was conducted but the fatigue strength could not be improved as expected. Depending on the individual case, fatigue failure due to shear occurred from the flange-shaped root of the nut. The improvement in fatigue strength by this method is only 1–2

kgf/mm². One reason is as follows. Except in special cases, fatigue failure initiates at the bolt. If a measure is taken for the nut, therefore, the fatigue strength cannot be improved as expected. (This method is just like scratching an itch from outside the shoe.) As there are several factors, other than uneven load sharing, which govern the fatigue strength of bolts, the measures proposed for nuts are effective for only one of the factors. Moreover, the bolts shown in Figure 4.103(b) and (c) are so complicated in shape that they are unsuitable for mass production and their application becomes somewhat difficult.

(c) Measures for bolts

There are only a few examples of measures taken for improvement of the fatigue strength of bolts. One such example is shown in Figure 4.104 [19]. The diameter of the body of the bolt is smaller than the nominal diameter. This type of bolt is called a bolt with reduced shank. If the diameter of the body is smaller than the nominal diameter over the entire bolt length, the bolt may be set out of centre with respect to the axis of the bolt hole. In this condition, a bending load may be applied to the bolt in addition to a tensile load. If a bending load is applied to the bolt, the fatigue strength of the bolt is decreased [17], and therefore it becomes necessary to prevent this decrease in strength. The reason why the measure shown in Figure 4.104(d) is more desirable than that shown in Figure 4.104(c) is that the stress concentration at the incomplete thread and the underhead fillet can be decreased.

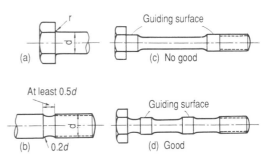

Figure 4.104 Conventional methods for improving the fatigue strength of bolts ((d) better than (c))

The effect of the bolt with a reduced shank is described below. Normally, bolts are used in a tightened condition. To use this type of bolt under the most favourable condition from the standpoint of fatigue strength, it is necessary to understand the relation between the external force applied to the fastening screw and the internal forces. The internal force means the load to be shared by the bolt and the fastened part, depending on the applied external force, but does not mean the stress.

Figure 4.105 shows the external force applied to the

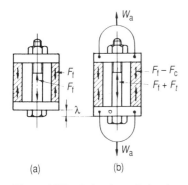

(a)　　　　　(b)

Figure 4.105 Balancing relation between external force and internal forces applied to the screw-fastened member (W_a, external force; F_f, F_t, F_c, internal forces)

bolt, nut and fastened part and the internal forces [17]. Figure 4.105(a) shows the condition in which the tensile force F_f induced in the bolt shank is balanced with the compressive force F_f induced in the fastened part when the bolt is tightened. Assume that an external force W_a is applied to the screw fastening. In this case, a tensile internal force F_t is applied to the bolt shank and a compressive force F_c is lost from the fastened part as shown in Figure 4.105(b). Under this condition, the fastening length is increased by λ. Let K_t be the tension spring constant of the fastening screw (load per unit elongation) and K_c the compression spring constant of the fastened part (load per unit contraction). In this case, F_t and F_c are expressed as shown below:

$$F_t = K_t \lambda \qquad (4.36)$$
$$F_c = K_c \lambda \qquad (4.37)$$

From the balance of forces

$$W_a = (F_f + F_t) - (F_f - F_c) = F_t + F_c \qquad (4.38)$$

Substituting equations (4.36) and (4.37) into equation (4.38):

$$W_a = (K_t + K_c)\lambda$$

Therefore

$$\lambda = \frac{1}{K_t + K_c} W_a \qquad (4.39)$$

Substituting equation (4.39) into equations (4.36) and (4.37):

$$F_t = \frac{K_t}{K_t + K_c} W_a, \qquad F_c = \frac{K_t}{K_t + K_c} W_a \qquad (4.40)$$

The ratio of the tensile force K_t added to the bolt by the external force W_a to the external force W_a applied to the fastening screw is expressed in terms of the internal force coefficient ϕ of the bolt as shown below:

$$\phi = \frac{K_t}{W_a} = \frac{K_t}{K_t + K_c} \qquad (4.41)$$

F_t and F_c in equation (4.40) can be expressed as shown below by using ϕ:

$$\left. \begin{array}{l} F_t = \phi W_a \\ F_c = (1 - \phi) W_a \end{array} \right\} \qquad (4.42)$$

The relation shown above is illustrated in Figure 4.106.

The relation between the force applied to the bolt and the fastened part and the elongation of the bolt (contraction of the fastened part) in the case where an external force is applied to the fastening screw is shown in Figure 4.106 with load as ordinate and elongation of bolt (contraction of fastened part) as abscissa [10,11,17]. This diagram is always introduced when the force applied to the bolt, particularly the fatigue strength of the bolt, is discussed. As it is not easy to understand this diagram in spite of its simplicity, a brief explanation is given below.

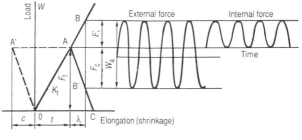

K_t:　spring constant of bolt
K_c　spring constant of fastened member
W_a　external force less than tightening force
F_f　tightening force to bolt
F_t　internal force due to external varying force W_a
Internal force due to external force W_a:

$$F_t = \frac{K_t}{K_t + K_c} W_a$$

Maximum force applied to bolt:

$$F_{max} = F_f \frac{K_t}{K_t + K_c} W_a$$

Figure 4.106 Relation between the force and shrinkage in the bolt and fastened member

The condition in which a tensile force F_t is being applied to the bolt is indicated by the point A. The relation between the force applied to the bolt and the elongation of the bolt is indicated by the line OAB. However, the line AB'C indicates the relation between the force applied to the fastened part and the contraction of this part. When an external force W_a is applied during the application of a tensile force F_f to the bolt, the elongation of the bolt, λ by the force W_a is equal to the elongation of the fastened part, δ, so far as the fastening between the bolt and the part is tight. Accordingly, both the bolt and the fastened part are elongated until the length of the line BB' becomes equal to the external force W_a. The forces to be shared by the bolt and the fastened part when the external force W_a is applied can be easily explained by following them from the start of fastening. The force to be shared by the bolt, F_t, and the force to be shared by the fastened part, F_c, are as shown by the stress wave form in Figure 4.105. If the external force W_a is increased until F_c becomes equal to F_f, the force by which the bolt is fastening the part becomes zero, and therefore the screw fastening becomes loose. If the screw fastening becomes loose, the whole external force W_a is applied to the bolt. To prevent loosening of the screw fastening by the application of the external force W_a, the following equation must be satisfied:

$$F_f \geq K_c W_a/(K_t + K_c) \tag{4.43}$$

Conversely, the screw fastening becomes loose if the external force W_a expressed by the following equation is applied to the bolt being tightened by the force F_f:

$$W_a \geq \left(1 + \frac{K_t}{K_c}\right) F_f \tag{4.44}$$

When the bolt is tightened to such an extent that it is not loosened by the external force W_a, the amplitude of the load to be borne by the bolt decreases with decreasing spring constant K_t of the bolt so far as the external force W_a and the compression spring constant of the fastened part, K_c, are constant. In other words, the force F_t to be borne by the bolt which is more likely to elongate within the elastic region is smaller with respect to the same external force W_a. For example, K_t for a long or narrow bolt is smaller than for a short or thick bolt.

In the case of a fastening screw to which an external force W_a with constant amplitude is applied, the fluctuating load to be borne by the bolt decreases with decreasing spring constant of the bolt and increasing spring constant of the fastened part. This is very advantageous from the standpoint of fatigue.

Figure 4.107 shows the relation between the internal force coefficient ϕ and l_f/d (fastening length/outside diameter of thread) [45]. When conventional bolts are used, the internal force coefficient ϕ is 0.1–0.3. As described previously (see Figure 4.67), the fatigue limit

of the bolt itself is 5–6 kgf/mm². Tolerable stress variations corresponding to variations in external force are 15–60 kgf/mm² according to equation (4.42). Accordingly, there may be cases where the fatigue limit of a fastening screw becomes higher than that of a welded structure.

On the basis of the results described above, this type of bolt is very effective as it does not receive the whole external force. Accordingly, the bolt has been used as a connecting rod bolt in engine casings and the like.

The fatigue strength of the bolt is hardly affected by the mean stress (see Figure 4.68) and the bolt does not become loose (the external force W_a does not satisfy equation (4.44) at all times). However, the bolt cannot be used where variations of external force W_a are directly applied to the bolt (for example, in a piston rod screw and hanging bolt).

(d) Fatigue strength of a fastening screw

The actual condition of use of the bolt with reduced shank which is shown in Figure 4.104(d) is described below. Figure 4.108 shows the S–N curve of a bolt with reduced shank in comparison with that of a bolt with nominal diameter body [26]. The outside diameter of the thread of both bolts is M10. In the case of the bolt with reduced shank, the pitch P is 1.5 mm and the area at the root in the transverse section, A_R, is 50.9 mm². In the case of the bolt with a nominal diameter body, the pitch P is 1.25 mm and the area at the root of the transverse section, A_R, is 55.0 mm². The mean test load P_m is 3000 kgf and the frequency rate is 1800 cycles/min. All stresses are arranged in relation to the nominal stress at the root diameter. It seems that the bolt with reduced shank is slightly higher in fatigue strength than the bolt with nominal diameter body, although we cannot say for sure because of some

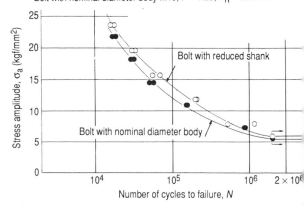

Mean load P_m = 3000 kgf
Bolt with reduced shank M10, P = 1.5, A_R = 50.9 mm²
Bolt with nominal diameter body M10, P = 1.25, A_R = 55.0 mm²

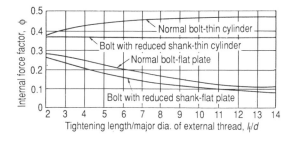

Figure 4.107 Quick calculation diagram of internal force factor for a typical fastened screw (for $K_t = K_c$)

Figure 4.108 S–N curve for a bolt with reduced shank (compared with that of a bolt with nominal diameter body)

differences in the condition of use between the two bolts in the strict sense. This can also be estimated from the flow of stress shown in Figure 4.103. In any case, the fatigue limit of both bolts is 5–6 kgf/mm². This value agrees with the value shown in Figure 4.67. Failures of all bolts initiated at the end face of the nut.

Figure 4.109 shows the *S–N* curve of a fastening screw (H. Kajino, M. Ide and S. Kondo, 1982, private communication) with dynamic external force W_a as the ordinate and the number of cycles as the abscissa. The frequency is 1800 cycles/min and the spring constants of the bolt with reduced shank and the bolt with nominal diameter body are 1.35×10^4 and 1.77×10^4 kgf/mm², respectively. The compression spring constant of the fastened part is calculated as 2.46×10^4 kgf/mm² [18]. The internal force coefficients ϕ of the bolt with reduced shank and the bolt with nominal diameter body are calculated as 0.355 and 0.419, respectively, from equation (4.40). As shown in Figure 4.109, the fatigue strength of the part fastened with the bolt with reduced shank (with smaller internal force coefficient ϕ) is larger than that of the part fastened with the bolt with nominal diameter body. All failures occurred at the bolt thread at the end section of the nut. For these reasons, extensive studies must have been made on the prevention of looseness to improve the fatigue strength of the bolt.

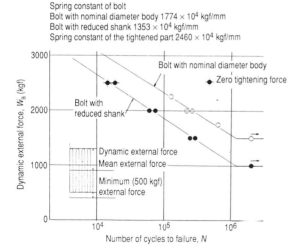

Figure 4.109 *S–N* curve for fastening screws (comparison between a bolt with reduced shank and a bolt with nominal diameter body)

The fatigue strength in the case of fastening with a bolt with reduced shank in the elastic range is shown in Figure 4.110 compared with that in the case of fastening with the same bolt in the plastic range (H. Kajino, M. Ide and S. Kondo, 1982, private communication). The test was conducted by the same method

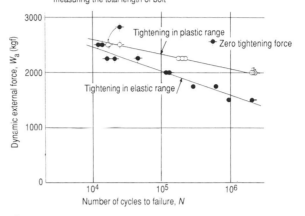

Figure 4.110 *S–N* curves for fastening screws (for fastening in the plastic range)

as that shown in Figure 4.109, except for the type of load cell used. Accordingly, the compression spring constant of the fastened part is calculated as 4.56×10^4 kgf/mm². In the case of fastening in the elastic range, the axial tension is set at '$0.6 \times$ yield load $= 3400$ kgf'. In the case of fastening in the plastic range, the spring constant was checked by measuring the overall length of the bolt. Fastening in the plastic range improves the fatigue strength by 43% (at 2×10^6 cycles), i.e. from 1400 kgf to 2000 kgf in terms of the dynamic external force. The probable reasons are that the compressive residual stress is developed by cold working of the thread root of the bolt and the loads to be shared by the bolt threads as shown in Table 4.19 and Figure 4.99 are changed. Yet another reason is that localized loading due to very small errors in pitch processing between the bolt and the nut is decreased by plastic deformation of the bolt threads (see Section 4.7.6).

The results of a study on the effect of prestressing which was conducted with the bolt made of SNCM630 (root diameter: 25 mm) are described below (H. Kajino, M. Ide and S. Kondo, 1982, private communication). In this study, the fatigue strength of the bolt was examined by applying a mean stress σ_m of 18 kgf/mm² instead of the bolt tightening force. As a result, the fatigue limit of 6.0 kgf/mm² was increased to 9.0 kgf/mm² by the application of a prestress of 43 kgf/mm². In other words, the fatigue limit of the bolt is increased by 50% by prestressing (see Figure 4.111). The difference in results shown in Figures 4.110 and 4.111 is attributed to the difference in the test methods employed. In the former method, the bolt was tightened under a load higher than the yield point and the

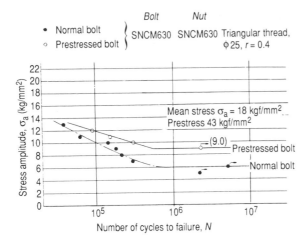

Figure 4.111   *S]N curves of prestressed bolt*

fatigue test of the fastening screw was conducted under this condition. In the latter method, however, the fatigue test was conducted by applying the specified mean stress after static yielding of the bolt in tension and load removal. Although the mean stress applied to the bolt in the former method differs from that in the latter method, both methods are based on what is nearly the same concept. Yunker pointed out that the fatigue strength is improved by tightening the bolt under a load higher than the yield point [28]. He attributed this improvement to the effect of compressive residual stress induced by the cold working of bolt

threads. The authors are of the opinion that the uniform load sharing among bolt threads and the decrease in localized one-side loading are also responsible for the improvement for the following reason. The effect of residual stress on fatigue strength is frequently likened to the effect of mean stress. Compared with the simple fatigue strength of a plain specimen, the fatigue strength of a bolt is little affected by the mean stress. Accordingly, an increase in the fatigue limit of a bolt by 50% due to the effect of compressive residual stress only is considered questionable.

4.7.5   *Effect of thread profile and other factors on fatigue strength*

(a)   Introduction
Normally, the fatigue strength of a bolt is 5–6 kgf/mm^2. This strength is not very high. Since bolts are mostly used under the tightened condition, part of a variation in the external force is shared by the fastened part. Accordingly, the critical stress for bolts is 15–60 kgf/mm^2, which is equal to or higher than the fatigue limit of welded structures.

First, the effect of thread profile on fatigue characteristics is studied below from the standpoint of improvement of the fatigue strength of the bolt (S. Nishida and C. Urashima, 1982, unpublished data).

(b)   Materials used, test specimens and test method
The chemical composition of the materials used for the test are shown in Table 4.21. The steels used for the nut

Table 4.21   *Chemical composition of materials used* (wt %)

Kind of steel	C	Si	Mn	P	S	Ni	Cr	Mo
SCM440 (ϕ 455)	0.41	0.35	0.73	0.0013	0.020	0.08	1.02	0.21
SNCM630 (ϕ 470)	0.29	0.25	0.44	0.009	0.006	2.97	2.98	0.59
S20C (ϕ 40)	0.19	0.01	0.41	0.008	0.005	–	–	–

The numbers in parentheses denote the size of material.

Table 4.22   *Mechanical properties*

Kind of steel	$\sigma_{0.2}$ (kgf/mm^2)	σ_B (kgf/mm^2)	El (%)	ϕ (%)	Impact value (kgf m)	
					$_vE$ 20°C	$_uE$ 20°C
SCM440	59.4	80.7	21.0	–	2.7	–
SNCM630	91.0	103.0	22.0	61.4	–	10.3
S20C	>25*	>41*	>28*	$H_v(10)$ 203	–	–

* Specified value.
$\sigma_{0.2}$, proof stress; σ_B, tensile strength; El, elongation; ϕ, reduction in area.

and bolt are SCM440 and SNCM630. In addition, S20C was used to study the effect of partial damage of the nut on the fatigue strength (see Table 4.25). All specimens were taken at a depth of 200 mm from the surface of the bar in such a way that the longitudinal direction of the bar becomes the central axis of the test specimen. The specimens, including the threads, were finished by turning. The mechanical properties of the steels are shown in Table 4.22. Tests were conducted on the following four items.

(i) Effect of type of thread

The effects of triangular thread, trapezoidal thread, positive buttress thread and negative buttress thread were studied. The profiles of the threads are outlined in Figure 4.112. The details of the threads are shown in Table 4.23. Although the outside diameter of the thread or the root radius may differ slightly from one type to another, the root diameter of all the threads is 25 mm.

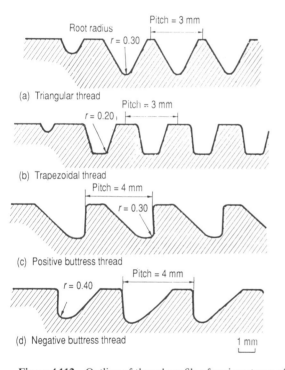

(a) Triangular thread

(b) Trapezoidal thread

(c) Positive buttress thread

(d) Negative buttress thread

1 mm

Figure 4.112 Outline of thread profile of various types of thread: (a) triangular; (b) trapezoidal; (c) positive buttress; (d) negative buttress

(ii) Effects of root radius

Three root radii r – 0.30, 0.50 and 0.70 mm – were selected. The outline and details of thread profiles are shown in Figure 4.113 and Table 4.24, respectively.

Table 4.23 Details of screw thread of bolt (see Figure 4.112)

Kind of steel	Type of thread	Details of thread profile
Bolt SCM440 Nut SCM440	Triangular thread ($\phi25$, $r^* = 0.30$)	Nut Pitch 3.0 60° 30° 30° 0.3 Bolt $\phi26.998$ $\phi25$ 1.9735 2.589
	Trapezoidal thread ($\phi25$, $r^* = 0.20$)	Pitch 3.0 Nut 30° 0.1 15° 15° 0.25 Bolt 5.596 2 $\phi25$ $\phi27$
	Negative buttress thread ($\phi25$, $r^* = 0.30$)	Pitch 4.0 1.2 Nut 0.35 0.2 2 Bolt 4.0 $\phi25$ $\phi27.4$
	Positive buttress thread ($\phi25$, $r^* = 0.40$)	Pitch 4.0 1.2 Nut 0.2 45° 0.3 2 Bolt 4.0 $\phi25$ $\phi27.4$

* = measured valve

(a) Root radius $r = 0.30$ mm

(b) Root radius $r = 0.50$ mm

(c) Root radius $r = 0.70$ mm

1 mm

Figure 4.113 Outline of thread profile with different root radii: (a) root radius $r = 0.3$ mm; (b) $r = 0.5$ mm; (c) $r = 0.7$ mm

Table 4.24 Details of bolt thread profile (see Figure 4.113)

Kind of steel	Type of thread	Details of thread profile
Bolt SCM440 Nut SCM440	Triangular thread (ϕ25, $r = 0.3$)	Nut Pitch 3.0 60° 30° 30° Bolt 0.3 ϕ26.998 ϕ25 1.974 2.598
	Triangular thread (ϕ25, $r = 0.5$)	Pitch 3.25 Nut 60° 30° 30° 0.15 0.5 Bolt ϕ26.615 ϕ25 1.890 2.815
	Triangular thread (ϕ25, $r = 0.7$)	Pitch 3.5 Nut 60° 30° 30° 0.7 0.15 Bolt ϕ26.431 ϕ25 1.907 3.031

(iii) *Effect of nut and bolt materials*

In almost all tests, a nut and bolt made of SCM440 were used. A study was also made with a material with higher strength (SNCM630) for the bolt. Moreover, as the nut is generally stronger than the bolt, a study was also made with a nut made of a material with lower strength (S20C).

(iv) *Effect of prestressing*

As described in the preceding section, this test method is effective but is not adopted widely on site. Accordingly, a brief description is given below.

Before a fatigue test, a tensile stress was statically applied to the nut and bolt in the axial direction.

Prestresses of 43 and 37 kgf/mm^2 were selected for the combination of nut and bolt made of SNCM630 and for the combination of bolt made of SNCM630 and nut made of S20C, respectively. All stresses are the nominal stresses at the root diameter (ϕ 25 mm). All specimens were subjected to partially tensile pulsating fatigue test with a mean stress σ_m. For the test, a servo type fatigue tester ($\pm$ 40 tf) was used. The frequency was 500 cycles/min. The *S–N* curves were obtained for all specimens. The fatigue strengths of the specimens at 2×10^6 cycles were compared. The testing condition is shown in Figure 4.114.

(c) **Results of tests and discussion**

Figure 4.115 shows the effect of root radius on the fatigue characteristic. In the case of the triangular thread which is the most widely used, its fatigue strength at 2×10^6 cycles (hereinafter the strength is called the fatigue limit unless otherwise specified) is 6 kgf/mm^2. (It is general practice to express the fatigue limit in terms of fatigue strength at 10^7 cycles. For convenience, however, the strength at 2×10^6 cycles may be used in view of the frequency of the fatigue tester.) As the specimens are finished by turning, a fatigue strength of 6 kgf/mm^2 is considered almost reasonable compared with the values shown in Tables 4.16 and 4.17 in Section 4.7.1. The fatigue limit of a positive buttress thread is nearly equal to this value, but the fatigue strength and fatigue limit of trapezoidal and negative buttress threads are slightly higher. This

Testing machine: electrical-servo-controlled
fatigue testing machine
Maximum capacity: $\pm$ 40 tf
Type of stress, partial pulsating tensile stress
Mean stress, constant at 18 kgf/mm^2
Frequency, 500 cycles/min

Figure 4.114   Testing condition

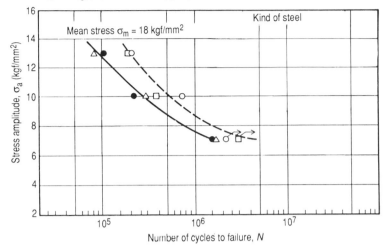

△   Triangular thread, φ 25, r = 0.30
□   Trapezoidal thread, φ 25, r = 0.20
●   Positive buttress thread, φ 25, r = 0.30
○   Negative buttress thread, φ 25, r = 0.40

Figure 4.115   Effect of kind of screw on fatigue strength

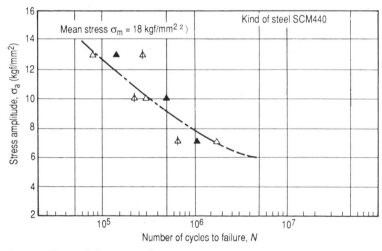

△   Triangular thread, φ 25, r = 0.30
◬   Triangular thread, φ 25, r = 0.50
▲   Triangular thread, φ 25, r = 0.70

Figure 4.116   Effect of root radius on fatigue strength

may be attributed to the relaxation of stress concentration at the root. Setting aside the trapezoidal thread, however, machining of a negative buttress thread is rather difficult. At the present time, this type of thread lacks wide applicability because of this very difficulty. In any case, the fatigue characteristic of a thread cannot be improved as expected even if the type of thread is changed. If workability is taken into account, the triangular thread has an excellent total balance beyond our expectation.

Figure 4.116 shows the effect of root radius. In this test, the root radius r was limited to 0.30–0.70 mm, and therefore the results of the test may not be applicable to all cases. From Figure 4.116 only, however, the root radius has little effect on fatigue strength, although some variations are observed. In all cases, the fatigue limit is 6 kgf/mm^2. The stress concentration at the root decreases with increasing root radius. However, if the root radius is increased, the rigidity of the threads is increased and localized contact with the internal threads is more likely to increase. It is considered that these effects cancel each other, causing little change in fatigue limit. It will be necessary to conduct further tests, changing the root radius over a very wide range. If the root radius is changed, the tensile strength of the bolt may be decreased. Accordingly, an improvement in fatigue strength only may produce an adverse effect. The thread profile should be determined by considering all the factors involved in a comprehensive way.

The effect of bolt material is shown in Figure 4.117. This figure shows the effect of changing the material of both nut and bolt from SCM440 to SNCM630. The tensile strength is increased by about 25% from 80 to 100 kgf/mm^2 by changing the material. In this case, there is little difference in fatigue strength between the two materials. However, the fatigue strength of SNCM630 is lower by about one-fifth in terms of the number of cycles.

It is widely known that the fatigue strength can be improved by increasing the tensile strength. However, the results shown in Figure 4.117 are the opposite. This effect can be explained as described below. That is, the effect shown in Figure 4.117 is attributed to two factors. One is that the bolt is a kind of notched specimen. In ordinary fatigue, the fatigue limit of a plain specimen tends to increase with increasing tensile strength. However, even if the tensile strength is increased with decreasing notch radius, this difference in tensile strength does not have a noticeable effect on the fatigue limit [31]. This is because the notch sensitivity of a notched specimen increases with increasing tensile strength and the fatigue strength of this

specimen decreases more than that of the plain specimen. The other is that the force is transmitted in the bolt through contact between the external threads and the internal threads. If the tensile strength is increased, the contact between the nut and the bolt is apt to become one-sided microscopically, although this contact seems to remain changed macroscopically. In other words, the effect shown in Figure 4.116 is attributable partly to localized contact. Examples in which the fatigue limit of the bolt is hardly changed even if the tensile strength is increased or in which the fatigue limit is markedly decreased with increasing bolt diameter [17,20] are attributed to the same factor. As bolts are manufactured separately from nuts, the pitch of the bolt differs from that of the nut in the strict sense even if the nominal pitch is the same. Moreover, the bolt receives tension in service but the nut is subjected to compression. Accordingly, the contact force applied to the bolt threads differs from that applied to the nut threads, causing localized contact. If the tensile strength of the material is high, localized contact is not relaxed as the material does not yield. If the tensile strength of the material is low, however, the material yields easily and the yielded part is plastically deformed, resulting in an increase in contact area. In other words, there may be cases of a reverse effect on the fatigue strength if the tensile strength of the material is increased.

Figure 4.118 shows the effect of the composition of a nut on the fatigue characteristic. The fatigue limit for S20C (7 kgf/mm^2) is improved by 17% compared with that for SNCM630 (6 kgf/mm^2). Furthermore, the

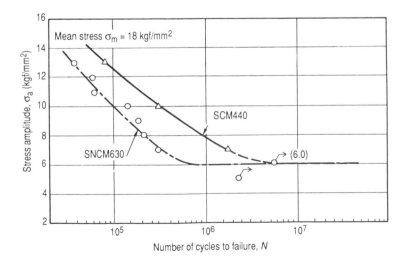

Figure 4.117   Effect of mechanical properties on fatigue strength

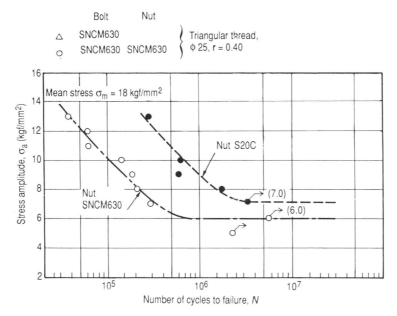

Figure 4.118 Effect of mechanical properties of nut on fatigue strength

fatigue strength is increased by about ten times in terms of the number of cycles. The nut has a larger root diameter than the bolt. The bolt is subjected to a tensile force at the time of loading, while the nut receives a compressive force and is hardly fractured. Accordingly, a considerable effect can be expected by making the nut of a material which is softer than the bolt material (see Section 4.7.6).

We often say that 'this easily fits or does not fit easily'. When a force is transmitted by the contact between more than two parts, such as in the combination of nut and bolt, the stress can be reduced if the force is received by a larger area. Normally, it is said that the microscopic contact area is less than 10% of the macroscopic contact area. If the nut or bolt or both are made of a softer material, the microscopic contact stress can be reduced. Since fracture is more likely to occur in the bolt because of tightening, better results will be obtained by using a soft material for the nut. According to the author's experience, it is desirable to set the ratio of tensile strength of the bolt to that of the nut within the range 1:0.4–0.8. Some researchers are of the opinion that the fatigue characteristic can be greatly improved by using a cast iron which has a lower elastic coefficient than carbon steel [21]. As described later, the use of cast iron aims at equalizing the load to be shared by the threads by decreasing the internal stress applied to the bolt by the external force.

Figure 4.119 shows the effect of prestressing on the fatigue characteristic (S. Nishida and C. Urashima, 1984, unpublished data). A prestress of 37 kgf/mm^2 was applied in the case of the combination of a bolt made of SNCM630 and a nut made of S20C. When both the nut and bolt are made of SNCM630, a prestress of 43 kgf/mm^2 was applied. In other words, the prestress to the former combination was lower by some 10% than that to the latter combination. In both combinations, the fatigue strength was increased by prestressing. The ratios of increase in fatigue strength were 38 and 50%, respectively. These increases are attributable to such factors as the improvement in the strength of bolt threads by cold working and the resultant effect of compressive residual stress, equal load sharing among the bolt threads, and the relaxation of microscopic localized contact between the bolt threads and the nut threads. In the case of the nut made of S20C shown in Figure 4.118, the nut is more likely to yield than the bolt. The reason is that the yielding at the bolt thread root was not sufficient compared with the case where the nut was made of SNCM630.

There are still several items requiring further study, including determination of the limit of improvement of fatigue strength by changing the prestress over a wide range, and selection of an optimum nut material in the case when prestress is applied. In any case, the fatigue limit of the bolt is improved by 50% by prestressing. Yunker pointed out that the fatigue strength is increased by tightening the bolt under a load higher than the yield point [28]. Maruyama explained the reason why the fatigue strength is improved by fastening in the plastic range [30]. Although the test method and effect differ in detail, both researches are based on the same idea. The results are summarized in Table 4.25.

Table 4.25 *Summarized fatigue test results for bolts*

Items for study	Type of bolt	Materials used		Size of bolt (mm)			Pretreatment	Test conditions		Fatigue strength at 2×10^6 cycles (kgf/mm²)	Increasing ratio of fatigue strength (%)	Remarks
		Bolt	Nut	Minor diameter	Root radius, r	Pitch		Mean stress (kgf/mm²)	Frequency (cycles/min)			
Fatigue strength of bolts with various threads	Triangular thread	SCM440	SCM440	25.0	0.30	3.0	No pretreatment	18.0	500	±6.0	100	Normal shape (standard bolt)
	Trapezoidal thread	SCM440	SCM440	25.0	0.25	3.0	ibid	18.0	500	±7.0	117	
	Positive buttress thread	SCM440	SCM440	25.0	0.35	4.0	ibid	18.0	500	±6.0	100	No effect
	Negative buttress thread	SCM440	SCM440	25.0	0.30	4.0	ibid	18.0	500	±7.0/±8.0	117/133	
Size effect	Triangular thread	SCM440	SCM440	32.0	0.80	3.6	ibid	18.0	350	(±6.0)	100	
	ibid	SCM440	SCM440	40.0	1.00	5.5	ibid	18.0	350	(±6.0)	100	
Effect of root radius r	ibid	SCM440	SCM440	25.0	0.50	3.25	ibid	18.0	500	±6.0	100	No effect
	ibid	SCM440	SCM440	25.0	0.70	3.5	ibid	18.0	500	±6.0	100	No effect
Effect of mechanical properties of bolt	ibid	SNCM630	SNCM630	25.0	0.40	3.25	ibid	18.0	500	±6.0	100	No effect†
Material of nut	ibid	SNCM630	S20C	25.0	0.40	3.25	ibid	18.0	500	±7.0	117	Slightly negative effect
Effect of prestressing (new method)	ibid	SNCM630	SNCM630	25.0	0.40	3.25	Prestress 43 kgf/mm²	18.0	500	±9.0	150	
	ibid	SNCM630	S20C	25.0	0.40	3.25	Prestress 37 kgf/mm²	18.0	500	±7.5	125	
	Gradual cut-off of bolt's thread*	SNCM630	SNCM630	25.0	0.40	3.25	Prestress 33 kgf/mm²	18.0	500	±12.0	200	CD bolt
Effect of gradual cut-off method of bolt's thread (new method)	Gradual cut-off of bolt's thread	SNCM630	SNCM630	25.0	0.40	3.25	Gradual cut-off from 1st thread to 8th one	18.0	500	±11.0	183	Engaged nut at 4th thread

* Gradual cut-off of bolt's thread means CD bolt.
† Fatigue life becomes about 1/10th.

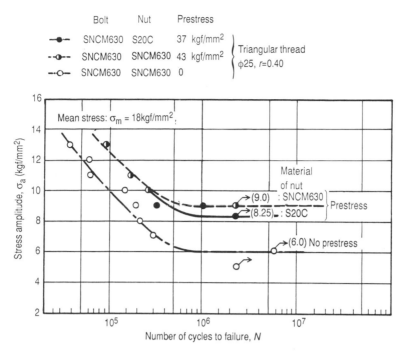

Figure 4.119 Effect of prestress on fatigue strength

4.7.6 A new method for improvement of the fatigue strength of bolts

(a) Shape of the CD bolt

CD bolt is the trade name of a bolt manufactured by the Nippon Steel Bolten Co. Ltd. CD stands for Critical Design for Fracture, which means the ultimate profile of a thread at the present time. This design is effective not only for fatigue properties but also for delayed fracture. The shaping method based on the CD bolt is called CD shaping. As know-how is involved in the use of the method, the manufacturer should be consulted. As described in Section 4.7.1, the fatigue strength of normal bolts is far lower than that of a specimen with a single notch which is made of the same material. This is because the load is transmitted in the bolt through the contact between the external and internal threads. The low fatigue strength is attributable to the following four factors:

1. Uneven load sharing among the threads
2. Concentration of tensile stress
3. Concentration of bending stress, and
4. Localized loading.

The factors, except factor (2), are not encountered with conventional notched specimens. Accordingly, these three factors are the dynamic phenomena which are characteristic of fastening screws. So far, only two methods have been proposed for the four factors as described in Section 4.7.4. One is the method for

ensuring equal load sharing among the threads. The other is the method in which the internal force coefficient is decreased depending on the part to be fastened as the bolts are normally used in a tightened condition (see Fig. 4.104). However, the former method is effective for only one of the four factors, while the latter method is effective only when the bolt is used by applying a tightening force. Moreover, this method becomes ineffective if the bolt becomes loose or if an external force which is large enough to cause plastic deformation of the bolt is applied.

The method being introduced below is a new method for improving the fatigue strength of bolts in which the fatigue strength of the bolt is improved markedly, regardless of whether a tightening force is applied to the bolt or not. The typical shape of the bolt (hereinafter referred to as the CD bolt) is shown in Figure 4.120. Shown here is the CD bolt with nominal diameter body. Needless to say, the idea of the CD bolt is also applicable to the bolt with a pitch diameter body and the bolt with a reduced shank.

A typical example of CD shaping is described below. In the case of coarse threads of normal size, the gradient of the part shaped by CD is about 6/100 and the incomplete thread is almost completely removed. The thread is connected to the body by a gentle arc ($R \geq 10$). The optimum end section of the nut is such that about 70% of the CD-shaped part of the bolt goes into the nut. If the distance between the end section of the nut and the incomplete thread is too long because

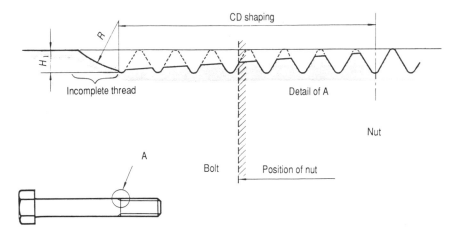

Figure 4.120   Typical shape of CD bolt

of a long thread, the length of the part from which the thread is removed should be increased. The fatigue strength at the end section of the nut and the incomplete thread can be considerably improved by this method. Moreover, the fatigue strength of the underhead fillet should be improved. For this improvement, the radius R under the head is increased ($R \geq 5\,\text{mm}$), or work hardening or compressive residual stress is applied to the rounded part under the head by cold working (shot peening or punching).

(b) General features of CD bolts and precautions in their use

The CD bolt has been developed from the dynamic standpoint to improve the fatigue characteristic of bolts. The CD bolt incorporates all the measures necessary for the improvement of the fatigue strength of bolts. Accordingly, there is a marked difference in effect between the CD bolt and conventional bolts. The typical shape of a CD bolt is shown in Figure 4.120. The fatigue characteristic of the CD bolt is excellent, as described below.

1. The load is shared uniformly.
2. As the height of the threads in engagement with the nut is reduced, the concentration of tensile stress is relaxed.
3. As the height of the thread is reduced, the distance between the loaded part of the thread and the thread root is decreased if the threads are assumed to be cantilevers. Accordingly, the concentration of bending stress at the root is relaxed.
4. As the contact surface pressure is increased, the contact area between the bolt and the nut is easily deformed and contact is made between the bolt thread and the tip of the thread on the nut side which is more likely to be deflected. Accordingly, localized loading due to low shaping accuracy or

other factors is reduced. In other words, the CD bolt is the bolt for which all measures necessary for the improvement of fatigue strength have been taken.

Precautions in the use of the CD bolt are described below. The optimum position of engagement between the CD bolt and the nut is determined depending on the characteristic thread profile of the bolt. On site, however, the nut and bolt are not necessarily engaged at the optimum position. In such cases, adjustment is made by changing the thickness of the washer. However, there may be cases in which adjustment with a washer is difficult. As a rough guide, the tolerance is about ± 1 thread. In other words, an allowance of about 5 mm can be made for tightening in the case of bolts with M20 coarse thread. This allowance is considered practically sufficient.

In cases where the bolts in stock are used in an emergency, the threads cannot always be engaged in an optimum way. The CD bolt is inconvenient on this point. In such cases, the use of a CD nut for which there is no limitation on the position of engagement is recommended, although the fatigue strength may be considerably decreased (see Figure 4.121). When a CD nut is used, the fatigue strength is improved only by about 30%, which is about one-third or one-quarter of the improvement made by the use of the CD bolt. This may be because measures for improvement are taken for only one of the general factors governing the fatigue strength of the bolt, i.e. uniform load sharing.

(c) Fatigue strength of the CD bolt

The fatigue strength of the CD bolt (at 2×10^6 cycles) is compared with that of other bolts in Table 4.25. As shown in this table, the fatigue strength of the CD bolt is nearly equal to that of a bolt with triangular thread. Conversely, the fatigue strength of a bolt with trian-

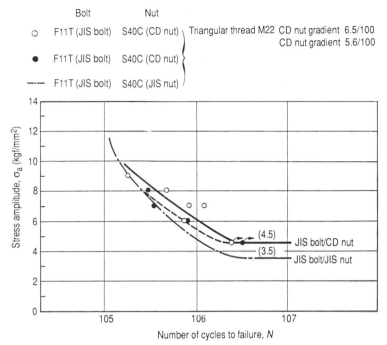

Figure 4.121   *S–N* curve for CD nut

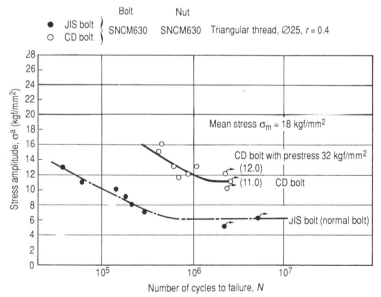

Figure 4.122   *S–N* curve for CD bolt

gular thread is not so low. It may be said that the triangular thread is nearly the ideal shape if its workability is taken into account. For this reason, we may be given the impression that the triangular thread is nearly the perfect thread, leaving little room for further improvement.

The fatigue characteristic curves in cases where CD shaping is applied are shown in Figures 4.122–4.125. The curves will be of reference in designing CD bolts. In the figures, the fatigue limit of normal bolts is not constant. This is because the fatigue strength of a bolt is dependent upon the combination with the nut. In the case of machine bolts, both nut and bolt are threaded by a lathe, and therefore this combination is nearly free

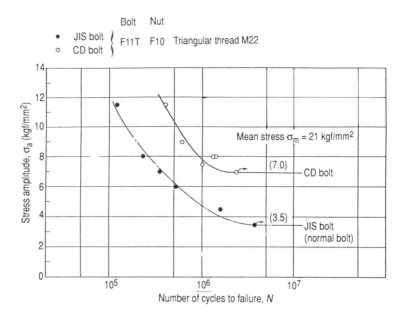

Figure 4.123   *S–N* curve for CD bolt for high-tension bolts

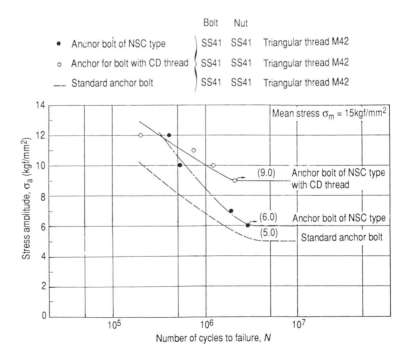

Figure 4.124   *S–N* curve for CD bolt for anchor bolts

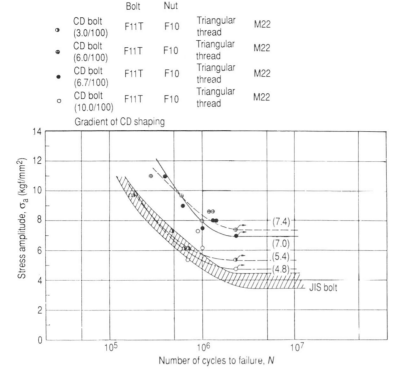

	Bolt	Nut		
○	CD bolt (3.0/100)	F11T	F10	Triangular thread	M22
◉	CD bolt (6.0/100)	F11T	F10	Triangular thread	M22
●	CD bolt (6.7/100)	F11T	F10	Triangular thread	M22
○	CD bolt (10.0/100)	F11T	F10	Triangular thread	M22

Gradient of CD shaping

Figure 4.125 Effect of gradient of CD shaping on the fatigue strength of CD bolt

from looseness (looseness means clearance). In the case of the high-tensile bolt and the anchor bolt, however, looseness is increased because of their uses (as the nut threads are over-tapped). If time allows, it is desirable to conduct a fatigue test of nuts and bolts made of various materials in various shapes by various methods and to design the bolt on the basis of the results of the tests.

For CD shaping, a gradient of about 6.0/100 is desirable from the standpoint of the fatigue characteristic. In the case of large-diameter bolts with fine or coarse threads larger than M40 in which the thread height is relatively lower for their diameter, it is necessary to change this gradient from 6.0/100. Ideally, it is necessary to eliminate or relax all the factors governing the fatigue strength of bolts as described previously. There are four factors. CD shaping is designed so that all these factors are eliminated only by taking a suitable measure for ensuring uniform load sharing. Accordingly, CD shaping should be done in line with this idea. In other words, the gradient of the part to be subjected to CD shaping can be so determined that the shaped part accounts for 60–70% of the whole thread and the thread engagement at the end section of the nut becomes 20–30% of the conventional engagement. These values are the empirical values.

(d) Static characteristic of the CD bolt

The static tensile characteristics of CD bolts, the proof loads of nuts and the torque coefficients of high-tensile bolts are shown in Tables 4.26, 4.27 and 4.28, respectively. As is apparent from these tables, there is little difference in static tensile characteristic between the CD bolt and the conventional bolt. Looking at the shape of the CD bolt shown in Figure 4.120, it may be intuitively apprehended that the threads may be fractured by shear or plastically deformed and the bolt may easily come out of the nut if the bolt is pulled. However, it will be understood from Tables 4.26 and 4.27 that such fears are groundless. The torque coefficient is given by the following equation:

$$k = \frac{T}{d \cdot N} \times 1000 \qquad (4.45)$$

where k is the torque coefficient, T is the torque (nut tightening moment, kgf m), d is the basic size of thread diameter of the bolt (mm) and N is the axial tension of the bolt (tensile force produced in the axial direction of the bolt, kgf).

The deformability of the bolt is shown in Figure 4.126. There is little difference in deformability between the CD bolt and the conventional bolt.

The results of tests on loosening due to vibration are

Table 4.26   *Static tensile properties*

Nominal diameter × total length (mm)	Type of bolt	$\sigma_{0.2}$ (kgf/mm²)	σ_B (kgf/mm²)	El (%)	ϕ (%)	Tensile strength of bolt with nut σ_B' (kg)
M16 × 55	CD bolt	110	116	18	68	18 000
	JIS bolt					18 500
M22 × 140	CD bolt	112	118	17	63	35 100
	JIS bolt					35 800

$\sigma_{0.2}$, proof stress; σ_B, tensile strength; El, elongation; ϕ, reduction in area.

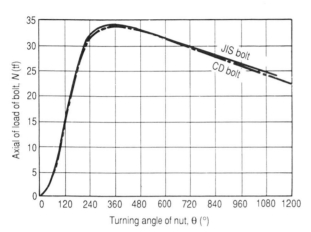

Figure 4.126   Deformability of bolt

Table 4.27   *Guarantee test result of nut*

Nominal diameter × total length	Type of bolt	Maximum load (kgf)
M16 × 55	CD bolt	18 200
	JIS bolt	18 300
M22 × 65	CD bolt	36 400
	JIS bolt	36 200

JIS bolt: conventional bolt.

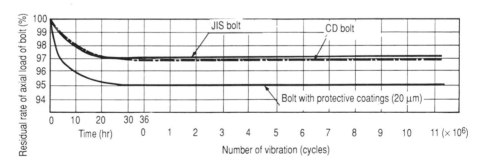

Figure 4.127   Test result for loosening of nut by vibration

Table 4.28   *Torque coefficient of high-tension bolt*

Nominal diameter × total length	Type of bolt	Torque coefficient, $\bar{K}$	Standard deviation, σ	Specification
M16 × 55	CD bolt	0.163	0.0046	B type $\bar{K}=0.150$–0.190
	JIS bolt	0.166	0.0012	≤ 0.013
M22 × 65	CD bolt	0.122	0.0088	A type $\bar{K}=0.110$–0.150
	JIS bolt	0.123	0.0045	≤ 0.010

JIS bolt: conventional bolt.

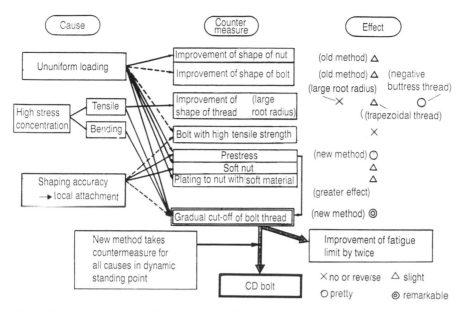

Figure 4.128   General factors governing the fatigue strength of bolts, countermeasures and their effects

shown in Figure 4.127, although the loosening does not belong to the static characteristic in the strict sense. There is little difference between the CD bolt and the conventional bolt.

The results described above are summarized in Figure 4.128. As the CD bolt, particularly the threaded part, has a higher fatigue strength, fracture of the CD bolt occurs under the head where the fatigue strength is considered to be high, while fracture of the conventional bolt initiates at the end section of the nut. An example of this type of fracture is shown in Figure 4.129.

(a)

(b)

Figure 4.129   Difference in position of failure between (a) CD bolt and (b) conventional bolt

References

1. Nishida, S., Urashima, C. and Masumoto, H. (1981) In *Proceedings of the 26th Symposium on Material Strength and Failure*, Sendai, Japan, p. 91
2. Nishida, S. (1981) *Fastening and Joining*, No. 33, Tokyo
3. Nishida, S. (1982) *Fastening and Joining*, No. 37, 1, Tokyo
4. Nishida, S. (1984) *Fastening and Joining*, No. 46, 1, Tokyo
5. Nishida, S. (1982) *Maintenance*, No. 30, 33, Tokyo
6. Nishida, S. (1982) *Maintenance*, No. 31, 40, Tokyo
7. Sasaki, T. (1984) *Machine Design*, **25**, 20, Tokyo
8. Muramatsu, S. (1980) *Journal of Japan Society of Mechanical Engineers*, **83**, 144, Tokyo
9. Yamana, M. (1972) *Last 30 Seconds – Investigation and Study of All Nippon Airways' Plane Crash off Haneda*, Asahi Newspaper Co. Ltd., Tokyo, p. 144
10. Kitsunai, Y. (1970) *Safety Engineering*, **9**, 249, Tokyo
11. Kitsunai, Y. (1974) *Safety Engineering*, **13**, 235, Tokyo
12. Kitsunai, Y. (1973) *Metallic Materials*, Vol. 13, Nikkan Kogyo Newspaper Co. Ltd., p. 32 Tokyo
13. Eto, G. (1972) *Metallic Materials*, Vol. 12, Nikkan Kogyo Newspaper Co. Ltd, Tokyo, p. 55

14. Kitsunai, Y. (1977) *Metallic Materials*, Vol. 17, Nikkan Kogyo Newspaper Co. Ltd., p. 31 Tokyo

15. Nagaoka, K. (1979) *Failure Analysis of Machine Parts*, Baifukan, Tokyo, p. 233

16. Pohl, E. J. (1964) *The Face of Metallic Fracture*, Vol. 2, Murich Reisensurance Co., p. 108

17. Machine Design Handbook Editing Committee (1973) *Machine Design Handbook*, Maruzen, Tokyo, p. 969

18. Japan Society of Mechanical Engineers (1961) *Design Data for Fatigue Strength of Metallic Materials*, p. 3

19. Ishibashi, T. (1954) *Prevention of Fatigue and Failure of Metals*, Yokendo, Tokyo, pp. 83, 220 and 256

20. Sunamoto, D. and Fujiwara, M. (1966) Technical Report of Mitsubishi Heavy Industries Ltd., **3**, 171, Tokyo

21. Ohtaki, H. *Machine Design*, **25**, 2, Tokyo

22. Yamamoto, A. (1975) *Theory and Calculation of Screw Fastening*, Yokendo, pp. 68 and 102, Tokyo

23. Sih, G. C. (1973) *Handbook of Stress Intensity Factors*, pp. 1, 4-2-1, Bethlehem, Pennsylvania, USA

24. Nishida, M. (1973) *Stress Concentration*, Morikita Shuppan, Tokyo, p. 666

25. Seike, S., Sakai, S. and Hosono, K. (1974) *Transactions of the Japan Society of Mechanical Engineers*, **40**, 140

26. Abe, H. (1967) *Journal of the Society of Steel Construction*, **2**, 55, Tokyo

27. Nishida, S., Urashima, C. and Masumoto, H. (1983) *Journal of the Iron and Steel Institute of Japan*, **69**, S708

28. Yunker, G. H. (1979) *Fastening and Joining*, No. 24, 12, Tokyo

29. Nisitani, H. and Nishida, S. (1970) *Journal of the Japanese Society for Strength and Fracture of Materials*, **5**, 84

30. Maruyama, K. (1984) *Fastening and Joining*, No. 44, 1, Tokyo

4.8 Environmental failure of fastening screws [1]

In Section 4.7, examples of the fatigue failure of bolts have been described. In the present section, examples of environmental failure are described. Fatigue failure of bolts occurs, regardless of steel grade, shape, etc. under repeated applications of some kind of stress. Environmental failure is limited to high-tensile bolts which are very sensitive to a corrosive environment. This section introduces examples of environmental failure of bolts to which repeated stress is apparently not applied or very small repeated stress is applied.

4.8.1 Failure of anchor bolts for a water-tube bridge

(a) Outline of failure

During his round of inspection, an inspector found broken bolts scattered on the (dry) ground. The bolts were assumed to be bolts used in a water-tube bridge. On closer inspection, it was found that they were anchor bolts used in the moving part of a bridge pier. Three out of four bolts were broken and their nuts were scattered. The bolts had been in service for 14 years and 9 months before their failure was detected (see Figures 4.130–4.132).

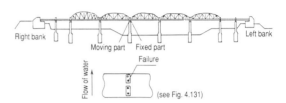

Figure 4.130   Condition of failure (see Figure 4.131)

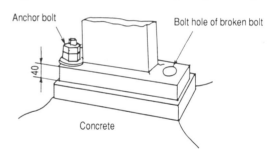

Figure 4.131   Anchor bolt set in place

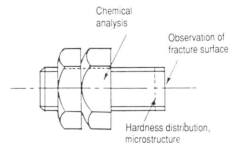

Figure 4.132   Sampling procedure

(b) Items investigated

1. Analysis of chemical composition and observation of the structure by optical microscope
2. Inspection of appearance
3. Mechanical properties (hardness distribution)
4. Observation of the fracture surface

(c) Results of investigation and discussion

(i) Appearance of the broken bolt

Figure 4.133 shows the appearance of the broken bolt. The failure initiated at a point about 47 mm from the end section of the nut. This distance is nearly equal to the sum of the thickness of the fastened part (40 mm) and the thickness of the washer (6 mm). The broken bolt was very corroded and the fracture surface was not clear because of rust. However, careful observation revealed a faint pattern extending in the a–b direction which looks like a crack propagation pattern (Figure 4.133(e)). (This direction is taken as the standard direction in Figure 4.13). The diameter in the standard direction is shorter by about 3 mm than that in a direction at right angles to the standard direction.

(a)

(b)

(c)

(d)

(e)

Figure 4.133 Outer appearance of broken bolt: (a) as received; (b) side appearance; (c) before pickling; (d) enlarged view of (c); (e) after pickling

Table 4.29   *Chemical composition* (wt %)

C	Si	Mn	P	S	Ni	Cr	Cu	Mo	B	Al
0.13	0.29	0.36	0.029	0.034	0.11	0.10	0.34	0.01	0.0001	0.004

(ii) Steel grade of broken bolt

The chemical composition and optical micrograph of the broken bolt are shown in Table 4.29 and Figure 4.134, respectively. The hardness distribution in the bolt is shown in Figure 4.135. On the basis of these data, the steel grade of the steel is judged to be equivalent to SS41. The structure is the typical ferrite–pearlite structure observed in low-carbon steels.

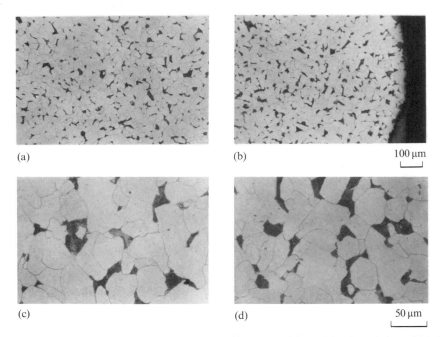

(a)   (b)   100 μm

(c)   (d)   50 μm

Figure 4.134   Microstructure of transverse section: (a) centre; (b) outer periphery; (c) enlarged view of (a); (d) enlarged view of (b)

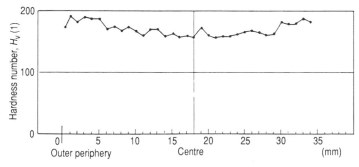

Figure 4.135   Hardness distribution (transverse section)

(iii) Observation of the fracture surface

The results of observation of the fracture surface are shown in Figures 4.136–4.138. As the bolt was exposed to a corrosive environment for a long period of time, an infinite number of corrosion pits is observed. It seems that changes were caused in the fracture surface due to the formation of a rust layer. Although not clearly seen, there is a pattern-like striation (Figure 4.138), and fissures are frequently observed in the fatigue fracture surface (Figure 4.137(b)).

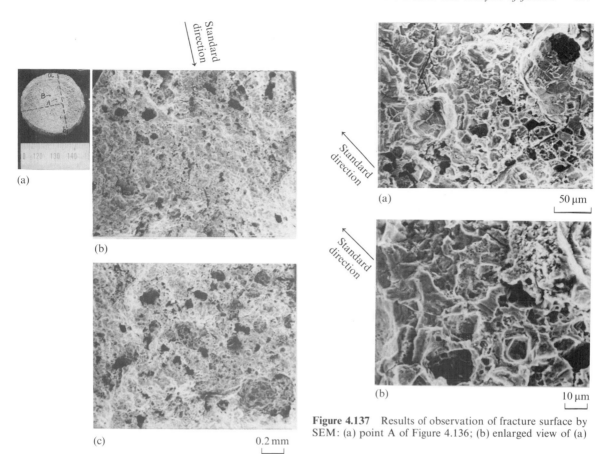

Figure 4.136 Results of observation of fracture surface by SEM: (a) appearance; (b) point A; (c) point B

Figure 4.137 Results of observation of fracture surface by SEM: (a) point A of Figure 4.136; (b) enlarged view of (a)

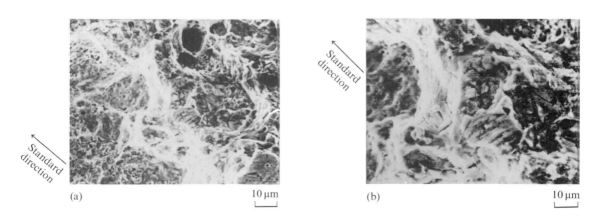

Figure 4.138 Results of observation of fracture surface by SEM: (a) point B of Figure 4.136; (b) enlarged view of (a)

(iv) Example of corrosion pits at the thread root

An example of this is shown in Figure 4.139.

(a)

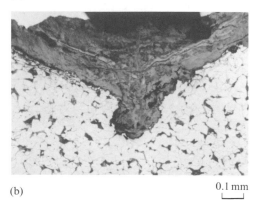

(b)

0.1 mm

Figure 4.139 Example of corrosion pits at the thread root: (a) before etching; (b) after etching with nital

(d) Summary

The bolt is made of a steel equivalent to SS41. No trace of macroscopic plastic deformation is observed on the fracture surface. The diameter in the direction which is supposed to be the direction of crack propagation is shorter by about 3 mm than the diameter in a direction at right angles to the above direction. The position of failure is the bottom end of the fastened part but not the end section of the nut where the maximum stress usually develops. A pattern-like striation and fissures are seen on the fracture surface, although not clearly. On the basis of these results, the cause of failure of the bolt is estimated to be fatigue failure caused by repeated application of the shear force produced by the expansion and contraction of the bridge pier during 15 years of service. Moreover, the corrosive atmosphere is also responsible. Accordingly, the diameter on the side where the shear force was applied is smaller by about 3 mm.

As a countermeasure, the anchor bolt hole is enlarged in consideration of the thermal expansion of the bridge pier, or the anchor bolt is set nearly at the centre of the bolt hole so that it is not affected by the thermal expansion of the bridge pier.

4.8.2 Failure of fastening bolt for the level gauge of an Ar-gas holder

(a) Outline of failure

Excessive leakage of Ar gas was detected from the rear side of the level gauge for an Ar-gas holder. On inspection, it was found that the fastening bolts of the level gauge were broken. Since many gauges of the same type were in use, the broken bolt was closely investigated to prevent recurrence of the same trouble. The gauge had been in service for 6 years and 10 months. Bolt details are: SUS630 (H900), 1/4-20UNC, unified thread, outside diameter: 6.35 mm, root diameter: 4.976 mm, pitch: 1.27 mm, thread length: 25.4 mm, overall underhead length: 63.5 mm, tightening torque: 2 kgf m. An axial tension of 317 kgf is produced when the gas pressure is 70 kgf/cm^2 (see Figure 4.140).

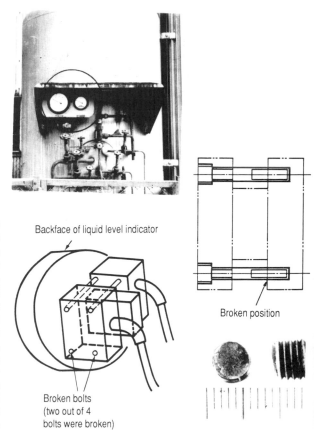

Backface of liquid level indicator

Broken bolts
(two out of 4
bolts were broken)

Broken position

Figure 4.140 (a) Outer appearance of level indicator; (b) position of failure of bolts; (c) position of failure and its fracture surface

(b) Items investigated

1. Analysis of chemical composition and observation of the structure by optical microscope
2. Inspection of appearance
3. Mechanical properties (hardness distribution)
4. Observation of the fracture surface and analysis of elements in the fracture surface

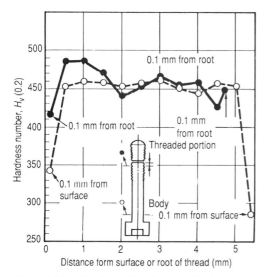

Figure 4.141 Hardness distribution of body threaded portion

(c) Results of investigation and discussion

(i) Bolt material

Table 4.30 shows the chemical composition and mechanical properties of the bolt. The Cu and Nb content deviate from the standard for SUS630.

The hardness distribution is shown in Figure 4.142. Though a noticable difference in hardness distribution is observed between the vicinity of the broken point and the body, the hardness of the whole bolt is very high.

The optical micrograph is shown in Figure 4.143. The structure is a tempered martensitic structure, but the tempering temperature is considered low.

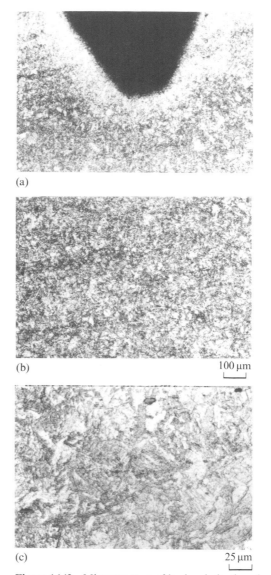

Figure 4.142 Microstructure of broken bolt: thread root of bolt; (b) centre of vertical section of threaded portion; (c) enlarged view of (b)

Table 4.30 *Chemical composition and mechanical properties of broken bolt*

| | Chemical composition (wt %) | | | | | | | | | Mechanical properties | |
	C	Si	Mn	P	S	Ni	Cr	Cu	Nb	PS (kgf/mm²)	TS (kgf/mm²)
Broken bolt	0.07	0.38	0.35	0.021	0.015	4.58	15.46	2.24	0.07	—	(150–160)*
SUS630 Specification	≤0.07	≤1.00	≤1.00	≤0.040	≤0.030	3.00/5.00	15.50/17.50	3.00/5.00	0.15/0.45	≤120	≤134

* Converted value from Vickers hardness number.
PS, proof stress; TS, tensile strength.

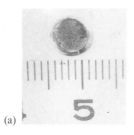

(a)

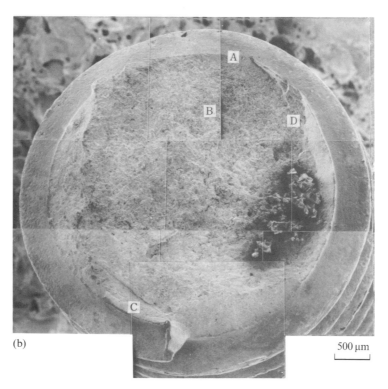

(b)

500 μm

Figure 4.143   (a) Outer appearance of fracture surface; (b) results of observation of fracture surface by SEM

(ii) *Observation of fracture surface and analysis of elements in the fracture surface*

It will be noted from Figures 4.144 and 4.145 that a large proportion of the fracture surfaces is the intergranular fracture surface. The final fracture is a ductile fracture. The existence of Cl and S was detected when part of the fracture surface was analyzed (see Figure 4.146). Corrosion pits were locally observed at the thread root and body of the bolt (Figure 4.147).

(iii) *Tightening force of bolt*

Assume that the bolt tightening torque is 2 kgf/m, the axial tension during service, 317 kgf, and the torque coefficient, 0.15. The stress to be created in the transverse section at the root under the above conditions is 123 kgf/mm^2. In general, the axial force for tightening a high-tension bolt is 75% of the standard yield point, and therefore the axial tension for tightening this bolt is 105 kgf/mm^2. It may be said that this tension is quite large. In cases where the tensile strength of a bolt is very high (about 150–160 kgf/mm^2) and the axial tension for tightening is large as described above, the bolt becomes sensitive to the influence of the environment and is apt to fracture in a brittle fashion (see Figure 4.147) [2,3].

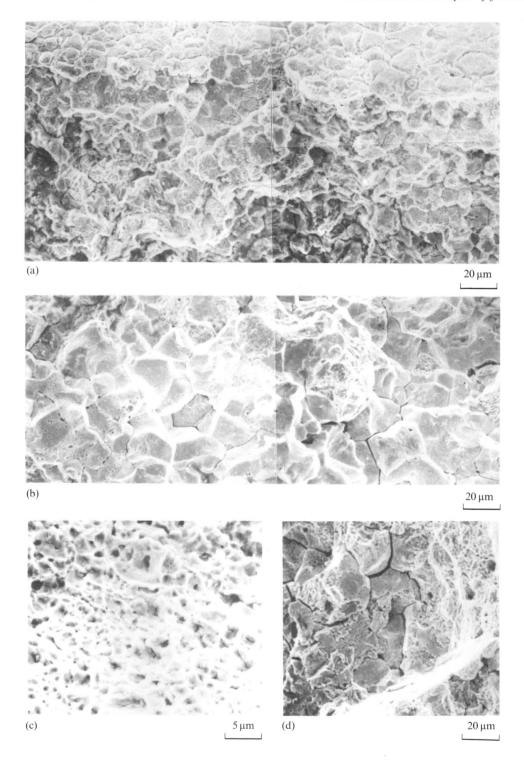

(a)

20 μm

(b)

20 μm

(c)

5 μm

(d)

20 μm

Figure 4.144   Results of observation of fracture surface by SEM (see Figure 4.123(b)): (a) enlarged view of (crack initiation) point A; (b) enlarged view of point B; (c) enlarged view of (final fracture) point C; (d) enlarged view of point D

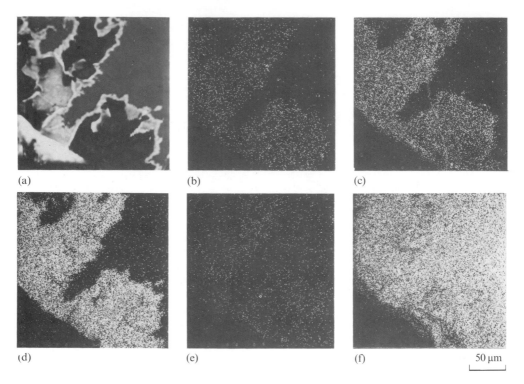

Figure 4.145 Chemical analysis of fracture surface: (a) absorbed electron image; (b) S; (c) Cl; (d) O; (e) Cu; (f) Cr

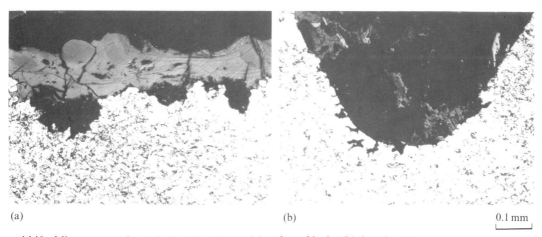

Figure 4.146 Microstructure in vertical section of bolt: (a) surface of body; (b) thread root

(d) Summary
Summarizing the results of the investigation described above, it is judged that the so-called stress corrosion cracking resulted from pitting caused in the high-tension bolt which was tightened under a large tightening axial force. To prevent this cracking, it will be effective to decrease the strength of the bolt by changing the heat-treating condition from H900 to H1075 and to decrease the bolt-tightening torque from 2 kgf m to 1 kgf m.

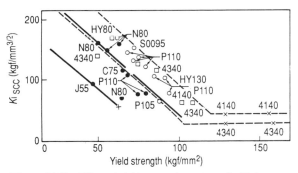

Figure 4.147   Effect of yield strength on K_{ISCC} in H_2S

4.8.3 Failure of fastening bolts for the small bell rod flange of a blast-furnace

(a) Outline of failure

All 20 fastening bolts for the small bell rod flange installed at the top of a blast-furnace broke after 6 months of service. Bolt details: steel grade: SCM435 (QT steel), M42 × 190 mm in length, thread length: 87 mm, stress applied to the bolt: repeated stress amplitude $\sigma_a = 2.9$ kgf/mm², mean stress $\sigma_m = 10.4$ kgf/mm². In addition, the bending stress was also applied (see Figure 4.148).

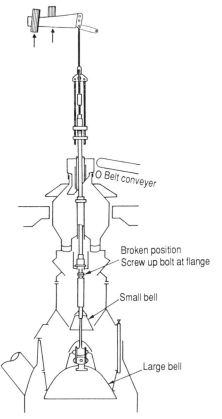

Figure 4.148   Schematic drawing of blast-furnace top position of failure in the bolt

(b) Items investigated

Nearly the same as those described in Sections 4.8.1 and 4.8.2.

(c) Results of investigation and discussion

(i) Appearance of the broken bolt

Out of 20 bolts, 18 bolts were broken under the head ($R = 0.7$ mm) and two bolts at the end section of the nut (although the number of broken bolts that could be collected was 15).

The surface of the broken bolts was very corroded. In some bolts, axial bending and cracks at the thread root were detected (see Figure 4.149).

Figure 4.149   Outer appearance of fracture surface of bolt

(ii) Chemical composition and hardness distribution

The chemical composition of the bolt is shown in Table 4.31. The composition satisfies the standard. The hardness distribution is shown in Figure 4.150. As there is a difference of 25% in hardness in the transverse section of the bolt, it is considered that the tempering of the bolt was insufficient, causing embrittlement of the surface.

Table 4.31   *Chemical composition* (wt %)

Remarks	Composition								
	C	*Si*	*Mn*	*P*	*S*	*Cu*	*Cr*	*Mo*	*Al*
Bolt A	0.36	0.25	0.79	0.027	0.013	0.03	1.00	0.18	0.054
Specification	0.32–0.39	0.15–0.35	0.55–0.90	≥ 0.030	≤ 0.030	≤ 0.030	0.85–1.25	0.15–0.35	–

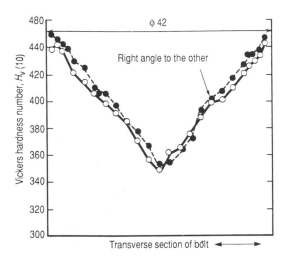

Figure 4.150   Hardness distribution in transverse section of bolt

(iii) Observation of the fracture surface

The results of observation are shown in Figures 4.149 and 4.151–4.155. At the failure initiation point in the fracture surface, transgranular failure and intergranular failure coexist, with the former type predominating. The transgranular fracture surface resembles the fatigue surface of high-strength steel but no striation is observed in it.

At a point 5–10 mm from the failure initiation point, transgranular failure and intergranular failure coexist but the proportion of intergranular failure is higher. Dimple and river patterns coexist in the final fracture surface.

(d) Summary

On the basis of the results described above, it is considered that fatigue cracks were caused by the stress concentration under the head of the bolt and their propagation was accelerated by the corrosive environment, finally resulting in failure.

To prevent failure, the following methods are suggested:

1. An increase in radius under the head ($R = 0.7 \rightarrow 5.0$ mm for M42 thread)

2. Equalization of axial tension of the bolt to an appropriate value by controlling the tightening torque
3. A decrease in surface hardness by tempering at 550–650°C, and
4. Rust-preventive treatment.

References

1. Nishida, S. and Urashima, C. (1984) Preprints of the 4th Environmental Failure Subcommittee of Fractography Committee, Society of Materials Science, Japan
2. Canter, C. S. (1973) *SCC and Hydrogen: Embrittlement of Iron Base Alloy*, National Association of Corrosion Engineers, USA
3. The Society of Steel Construction (1979) 'Final Report on Exposure Test by the Bolt Strength Group', *Journal of the Society of Steel Contruction*, **15**, 1

4.9 Failure of gears [1,2]

4.9.1 General configuration of failure

The gear and the bolt are typical machine elements. It is not too much to say that gears are used in almost all machines and equipment. Failure of gears not only obstructs normal operation of machines and equipment but also causes failure of related parts. Gears may fail if their design, fabrication and service conditions are improper. Close study of these factors is essential for the prevention of gear failures. Gear failures are complicated and cannot be classified simply. The classification of gear failures by the AGMA (American Gear Manufacturers Association) is well known. In Japan, gear failures are classified as shown in Table 4.32 [3].

Gear failures are roughly divided into the following two types.

1. Failure attributable to the strength of the gear material and the applied load, such as breakage, cracking and spalling
2. Failure attributable to an inability to form an oil film due to wear, seizing, etc. and improper contact of gears.

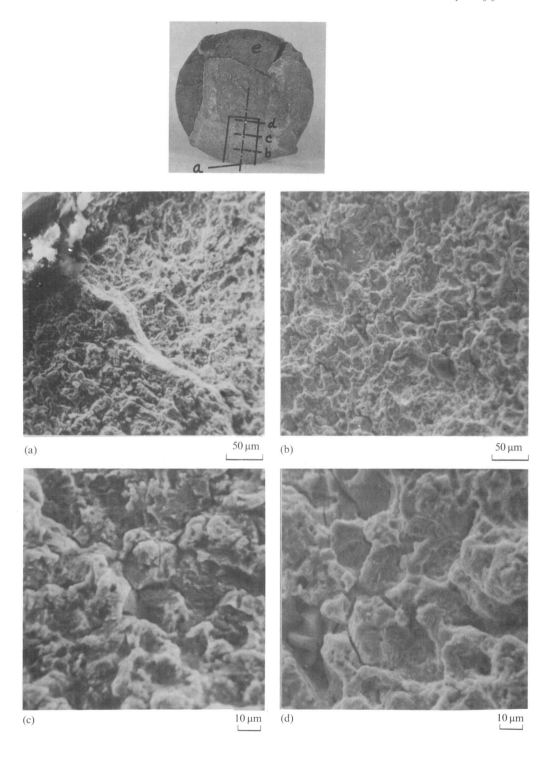

Figure 4.151   Results of observation by SEM of fracture surface of bolt A: (a) crack initiation point (point a); (b) 5 mm from initiation point (point b); (c) enlarged view of (a); (d) enlarged view of (b)

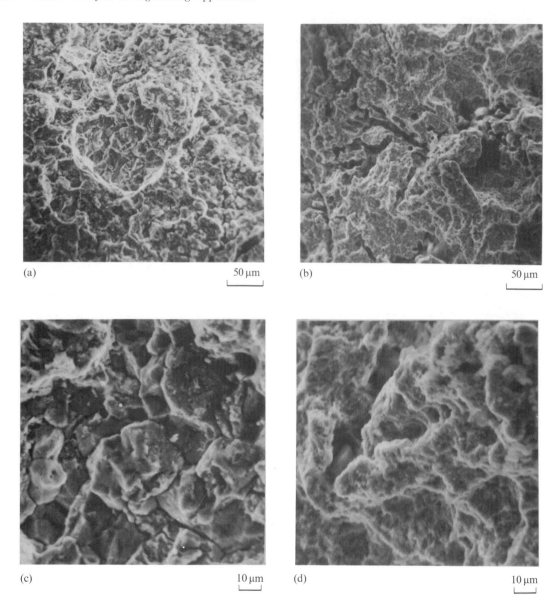

Figure 4.152   Results of observation by SEM of fracture surface of bolt A: (a) 10 mm from initiation point (point c); (b) 10 mm from initiation point (point c); (c) enlarged view of (a); (d) enlarged view of (b)

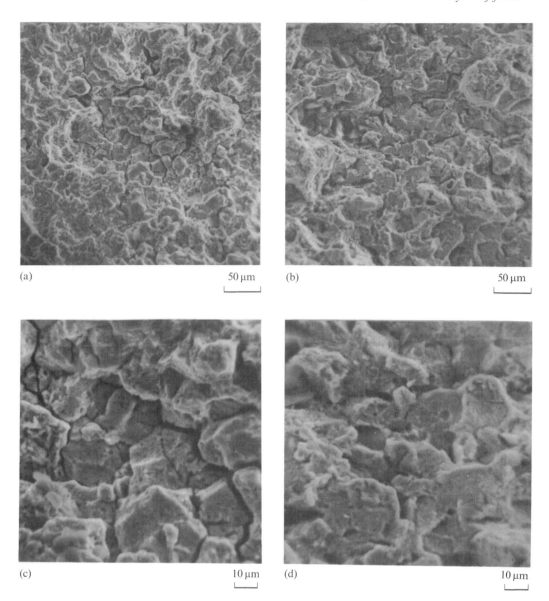

(a)   50 μm

(b)   50 μm

(c)   10 μm

(d)   10 μm

Figure 4.153   Results of observation by SEM of fracture surface of bolt A: (a) 15 mm from initiation point (point d); (b) final fracture surface (point e); (c) enlarged view of (a); (d) enlarged view of (b)

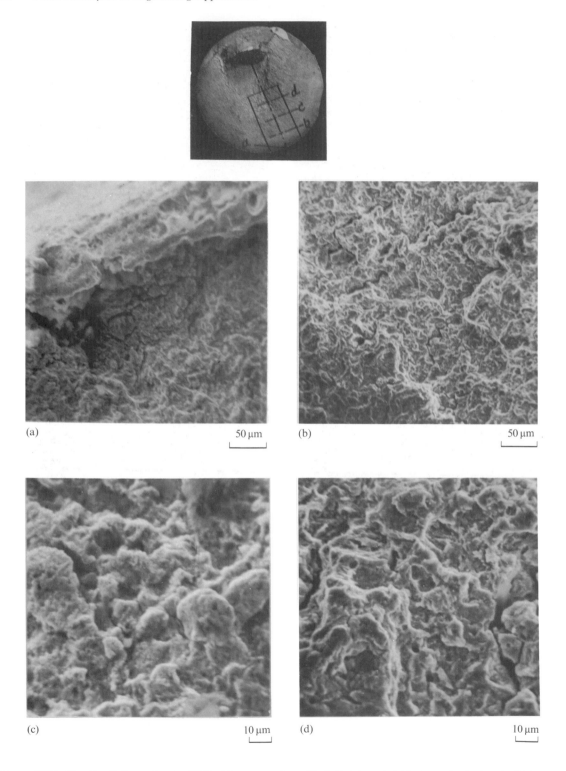

Figure 4.154 Results of observation by SEM of fracture surface of bolt B: (a) crack initiation point (point a); (b) 5 mm from initiation point (point b); (c) enlarged view of (a); (d) enlarged view of (b)

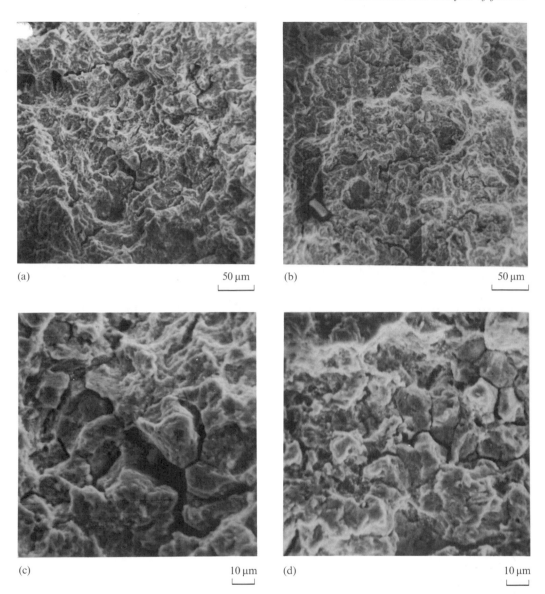

(a)　　　　　　　　　　50 μm

(b)　　　　　　　　　　50 μm

(c)　　　　　　　　　　10 μm

(d)　　　　　　　　　　10 μm

Figure 4.155　Results of observation by SEM of fracture surface of bolt B: (a) 10 mm from initiation point (point c); (b) 15 mm from initiation point (point d); (c) enlarged view of (a); (d) enlarged view of (b)

Table 4.32   *Classification of failure of gears*

Final failure	Progressive failure
1. Failure	1. Wear
(a) Fatigue failure	(a) Abrasive wear
(b) Overload failure	(b) Scratching
(c) Shear failure	(c) Fretting
2. Collapse	2. Plastic deformation
3. Melting	(a) Plastic flow
4. Jam	(b) Rippling
5. Abrasion	(c) Ridging
	(d) Declination of tooth
	3. Fatigue in surface of
	tooth
	(a) Pitting
	(b) Spalling
	4. Thermal cause
	(a) Scoring
	(b) Burning
	5. Other causes
	(a) Crack
	(b) Failure due to
	interference
	(c) Failure due to
	jamming foreign
	object
	(d) Corrosion
	(e) Electrolytic
	corrosion
	(f) Swelling

These failures are complicated as they are caused by a multiplier effect of heat, load, environment, etc. Normally, the majority of gear failures are associated with the strength of the gear material and load, such as breakage and spalling of teeth and pitching, rolling and peening of the toothed surfaces. The causes of failures include fatigue, overloading, improper heat treatment, the existence of tensile residual stress, etc. followed by causes resulting from insufficient lubrication and improper contact of tooth surfaces, such as abrasive wear by fine particles, scratching and scoring. Gear failures are also attributable to corrosive wear and corrosion fatigue due to the chemical impurities contained in the lubricating oil and use of the gears in a corrosive environment.

In many cases, gear failure is caused by the combined effects of stress, material, environment, etc. Accordingly, studies must be made on a wide variety of items, including design, fabrication, operation, maintenance and control. As the investigation of cause of failure is difficult, rich experience is required.

In this section, several examples of gear failure are introduced, describing the method and results of investigation, the causes of failures and countermeasures.

4.9.2 *Case 1: failure of teeth due to improper heat treatment*

(a) Outline of failure

During six years of service, all 13 teeth of the triaxial pinion of the reduction gear for traversing equipment were broken and one of 40 teeth of the treble gear was bent. The tooth surfaces of both pinion (SCM440) and gear (S45C) were subjected to induction hardening (see Figure 4.156).

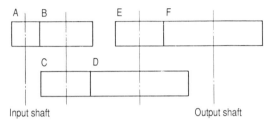

Pinion (E): All of 13 teeth were broken
   (SCM440, quenched by induction heating)
Gear (F): One of 40 teeth was bent
   (S45C, quenched by induction heating)

Figure 4.156   Position of failure of reduction gear

(b) Results of investigation and discussion

On inspection it was found that one of the broken pinion teeth which was supposed to be the tooth that fractured first was badly deformed as it was bitten by other teeth. Another twelve teeth were broken from the root fillet (the part where maximum stress is applied). Pitting had occurred on the contact surface of the teeth but contact was fairly good.

The macroscopic condition of the broken part is shown in Figure 4.157. Shown here is a tooth which

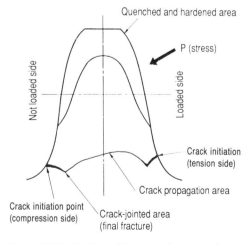

Figure 4.157   Outline of fracture of gear tooth

was not severely deformed. The failure initiated at the fillets on the tension and compression sides. The main crack propagating from the tension side joined the crack propagating from the compression side, resulting in final failure.

According to the results of microscopic observation of the fracture surface (SEM), the crack on the compression side was caused earlier than that on the tension side. The crack on the compression side propagated only by a short distance but the crack on the tension side propagated up to the crack on the compression side.

The condition of failure described above indicates that the failure is a fatigue failure of a non-hardened gear. As is apparent from Figure 4.157, the hardening range is observed from the vicinity of the pitch line to the tip of the tooth, but the hardening range is not present at the root.

It is said that cracking in a surface-hardened gear initiates on the tension side but not on the compression side. In the case of a non-hardened gear, however, the stress condition is as shown in Figure 4.158. The bending stress (a) and the shearing stress (c) are created by one component (out of two components of force) which is vertical to the centre line of the tooth, while the compressive stress (b) is inducted by the component of force which is acting in the direction of the centre line. This may be the reason why the absolute value of stress at the root on the compression side becomes greater than that on the tension side. It is generally said that the absolute value of stress on the compression side is greater by 20–30% than that on the tension side [4]. In addition, the residual tensile stress is created inside the root fillet on the compression side, although the stress has not yet been measured. It is considered that the stress is responsible for crack propagation to some distance. The residual tensile stress is attributable to partial yielding due to heat treatment of the tip of the tooth or an external force.

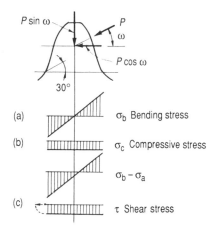

Figure 4.158   Stress condition at root of tooth

(c) Estimation of the cause, and countermeasures

On the basis of the results described above, it is considered that the cause of failure of the gear is the low fatigue limit attributable to improper induction hardening (non-hardening) at the root fillet.

As the failure condition of all the teeth of the triaxial pinion, except the excessively deformed one which must have been broken first, is nearly the same, it is estimated that all the teeth were broken at nearly the same time.

To prevent this type of failure, the hardening range should be checked by conducting an acceptance inspection, or induction hardening should be changed to total hardening.

4.9.3 Case 2: failure due to improper overloading and improper heat treatment

(a) Outline of failure

During a comparatively short period of service, the gear for a rocker shovel reversing mechanism was broken as shown in Figure 4.159. Two (adjoining)

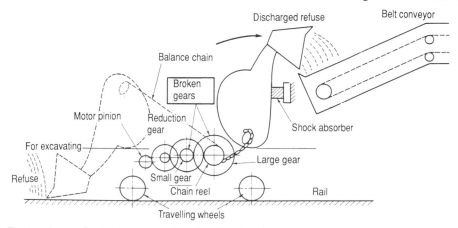

Figure 4.159   Fractured gears for turning mechanism of rocker shovel

Broken teeth

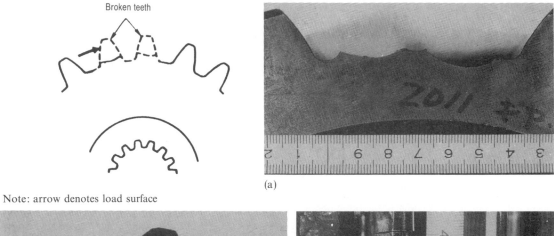

(a)

Note: arrow denotes load surface

(b)

(c)

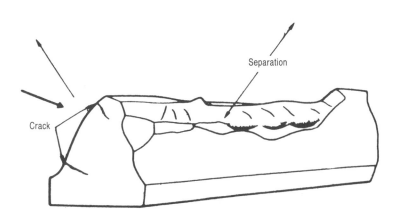

Separation

Crack

Figure 4.160   Outer appearance of broken gear: (a) broken at root of tooth of large gear (G6); (b) cracks at tooth root and tooth tip surface of small gear (G5); (c) separation at tooth tip surface

Table 4.33 *Main parameters of broken gear*

Gears	Module, m	No. of teeth, z	Amount of addendum modification, mx	Pitch circle diameter, d_0	Intermeshing pitch circle, d_b
G5	10	13	+5 (+0.5)	130	134.300
G6		32	+3.3 (+0.33)	320	330.586

Addendum circle diameter, d_k	Reduction ratio, i	Transmission torque, T (kgf m)	Tangential force, W (kgf)	Face width, b
158.286	18.587	422.9	6 506	90
344.886	(2.462)	1040.9		80

Addendum circle diameter, y	Bending stress σ_b (kgf/mm²)	Surface pressure σ_b (kgf/mm²)	Compensating bending stress σ_b' (kgf/mm²)*	Contact ratio, ε
0.292	24.8	141.8	17.1	1.27
0.382	21.3		17.0	

* Compensating bending stress: $\sigma_b' = \sigma_b (T_g'/T_g)^2$; T_g and T_g' are tooth thicknesses at base circle of standard gear and profile shift gear (addendum modification coefficient, x).

Table 4.34 *Material of broken gear**

	G5	G6
Material	SNC815	SCM420
Heat treatment	HCQ	HCQ
Yield point	80 kgf/mm²	80 kgf/mm²

* Caburizing and induction quenching.

teeth of the gear (gear G6) were broken from the root fillet and 11 out of 13 teeth of the pinion (pinion G5) were cracked at the tip and root of the tooth. Moreover, spalling occurred at the tip of three teeth (see Figure 4.160). The main specifications and materials of the broken gears are shown in Tables 4.33 and 4.34, respectively.

(b) Results of investigation and discussion

As these broken gears are used in the rocker shovel reversing mechanism, the loaded tooth surface receives a large reaction force during chain winding and at the end of winding (when the chain is stretched). It is also expected that loads in excess of the allowable stress are applied depending on the operating condition. Judging from the condition of failure, cracks initiated at the loaded surface in both gears. Cracks, spalling, crushing, deformation, etc. are found in the pinion. According to the results of microscopic observation of the fracture surfaces of the gear and pinion, the fracture surface of the gear is a low-cycle fatigue fracture surface. It seems that cracks in the pinion were caused by a comparatively large stress, but a high-cycle

fatigue fracture surface was produced in the crack propagation range.

The macroscopic cross section of both broken gears and the position where the microstructure was observed by optical microscope are shown in Figure 4.161. The results of observation of the structure of the gears are shown in Figure 4.162. The gear is in the overcarburized condition (see Figure 4.162(3) and (4)),

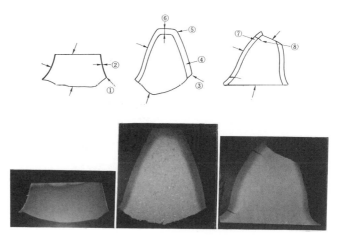

Figure 4.161 Microscopic cross section and position of microscopic observation of broken gear

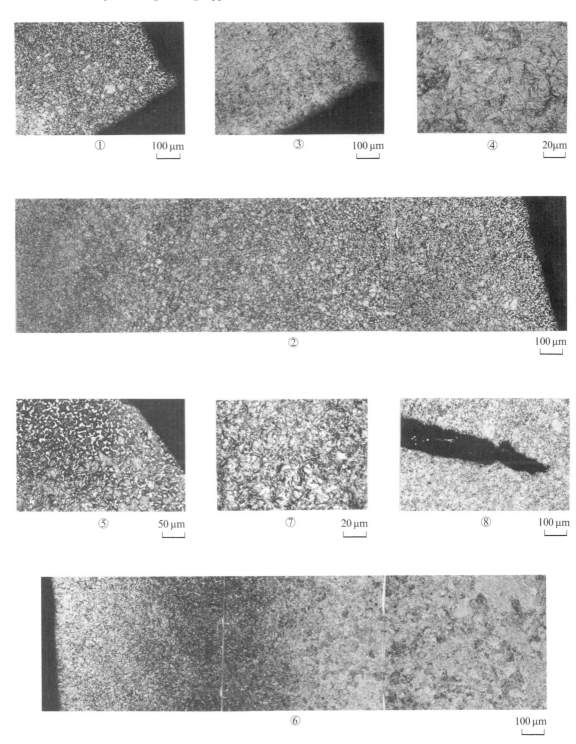

Figure 4.162   Results of observation of optical microstructure (the numbers cross refer to Figure 4.161)

and a free cementite structure and network cementite structure are observed. Moreover, uneven carburization and an uneven structure are also observed. No abnormal structures are observed in the pinion.

The results of hardness measurements at the end face of the gear are shown in Figures 4.163–4.165. As shown in Figures 4.163–4.165, the gear is in the overcarburized condition and the hardening range is quite wide in the cross section of the gear in particular.

Since the fatigue limit of a carburized and quenched specimen of SCM420 ($H_{RC}62$, H_v750, $\phi 9\,mm$) in rotational bending is about 70–80 kgf/mm², it is considered that stresses greater than this fatigue limit were applied to the gear. Since an overcarburized structure was observed, a decrease in toughness and deterioration of the material due to an increase in residual austenite are suspected. As the hardened layer of the gear is brittle, initial cracks that developed in the gear propagated rapidly, resulting in final failure.

(c) Estimation of cause and countermeasures

As the gear was overcarburized, it is considered that failure of the teeth was caused by the rapid propagation of cracks, and the broken piece of the teeth was bitten by the pinion, causing cracking and spalling in the pinion. To prevent this type of failure, it is necessary

1. To improve the strength of the gear by increasing the module or face width, and
2. To improve the material of the gear or the heat-treating method.

4.9.4 Case 3: Wear and spalling of teeth due to improper carburization and hardening

(a) Outline of failure

During about a year of service, the gear was started and stopped nine times and operated for about 7900 hours. The surfaces of all teeth on the drive side were worn (about 30 μm) and spalling resulted in one of the teeth. Moreover, damage, such as dents and breakage, were caused on the gear tooth surface. The specification and failure of the pinion of the turbine reduction gear are shown in Figure 4.166.

(b) Results of investigation and discussion

Macroscopic observation of the failure revealed many scored pits, plastic flow and fine cracks at the tip of the tooth. Dents were caused by the one-sided loading and abnormal wear at the root of the tooth (see Figure 4.16). Judging from this condition, failure is attributable to wear due to an increase in surface pressure and insufficient hardness of the tooth surface, as well as to scoring due to an increase in oil temperature and a lack of oil.

The chemical composition of the gear material is shown in Table 4.35, while the macroscopic structure

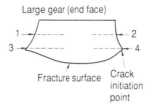

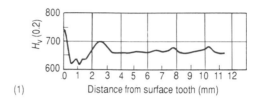
(1)

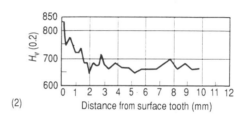
(2)

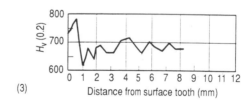
(3)

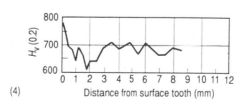
(4)

Figure 4.163 Hardness distribution at gear end face (at end face of large gear)

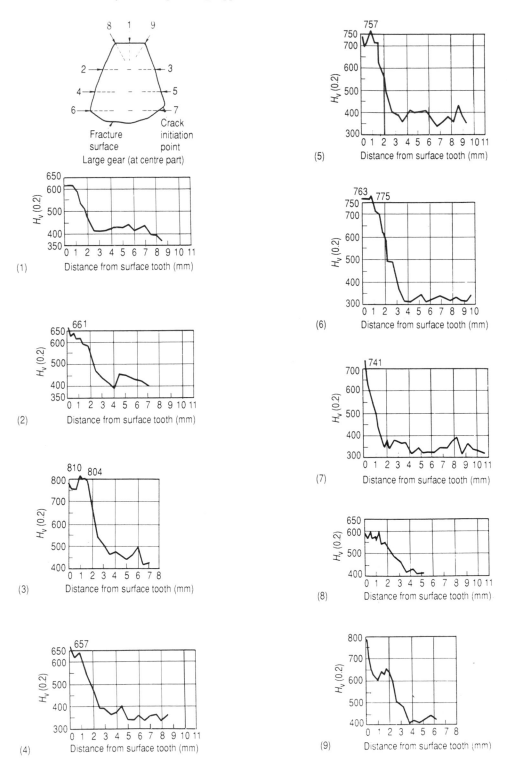

Figure 4.164   Hardness distribution at gear end face (at centre of large gear)

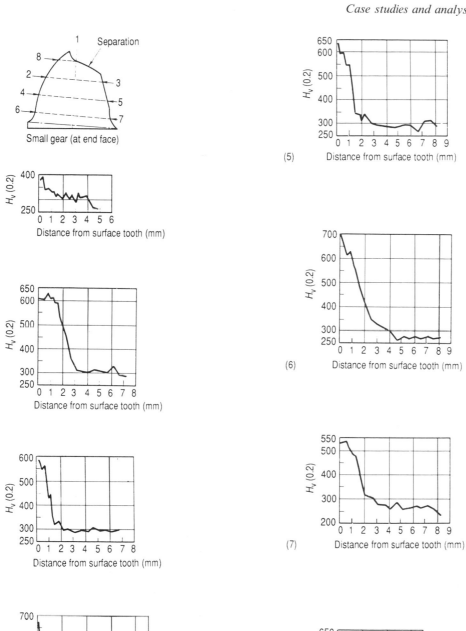

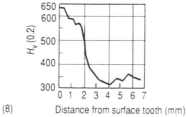

Figure 4.165   Hardness distribution at gear end face (at end face of small gear)

Table 4.35   *Chemical composition (wt %, SNC815)*

	C	Si	Mn	P	S	Ni	Cr
Specimen	0.15	0.28	0.47	0.016	0.011	3.16	0.77
Specification	0.12–0.18	0.15–0.35	0.35–0.65	<0.03	<0.03	3.00–3.5	0.7–1.0

No. of teeth 26
Module 6
Pressure angle 20°
Helix angle 16°56'33" (right)
Addendum circle diameter 170.889 mm
Hardness of tooth surface $\geq R_c 58$
Rotation frequency 8885 rev/min
Finishing method carburized quenching and grinding
Transmission power 5080 kW
Material SNC815

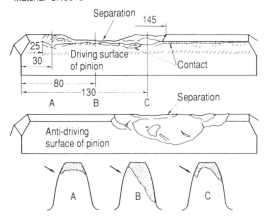

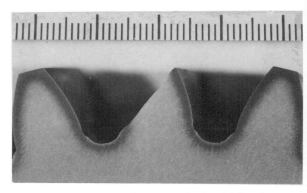

Figure 4.167   Macroscopic structure in the vicinity of broken tooth

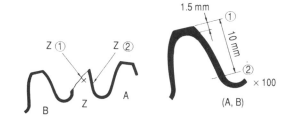

Figure 4.166   Specifications and outline of broken gear

Figure 4.168   Position of microscopic observation

Figure 4.169   Microscopic structure of gear (see Figure 4.168)

in the vicinity of the broken teeth is shown in Figure 4.167. The chemical composition satisfies the standard value. Macroscopically, the whole gear is heat-treated.

The position where the structure was observed by an optical microscope is shown in Figure 4.168. The results of observation are shown in Figure 4.169. In the vicinity of the teeth surface, phases with slightly low hardness (the sorbite phase and the ε phase in which nitrogen is in solution, $H_v = 258–480$, by X-ray diffraction) were observed. That is, the structure is uneven.

Contact of the tooth surface was then investigated. The results of observation of the tooth which was not macroscopically fractured are shown in Figure 4.170. In general, the gear receives the shear force applied in the tip and root directions with the pitch point as the centre. Accordingly, the fractional force applied to the tooth surface acts in the opposite direction with the pitch point as the centre. In the interior, the shear stress is applied in the direction opposite to the direction of fractional force to balance with the frictional force at the tooth surface (see Figure 4.168). From the pitch point to the root of the tooth, shear cracks are produced in the direction opposite to that of the frictional force. In the early stage of the operation, the cracks are produced in the direction opposite to that of the frictional force (in the direction of the root) and are also produced in the opposite direction as a result of shifting of the pitch point towards the tip of the tooth due to changes in contact during operation (see Figure 4.171).

The results of the mechanical test are shown in Table 4.36, while the design values are shown in Table 4.37. No specific problems are noted in the results of the mechanical test and design values.

(a)

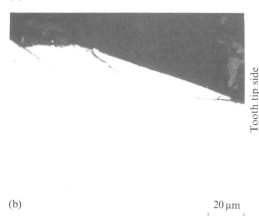

(b)

20 μm

Figure 4.170 Contact condition of gear tooth (not broken teeth; the pitch point is 6 mm from the tooth tip): (a) 5.5 mm from tooth tip; (b) 4 mm from tooth tip

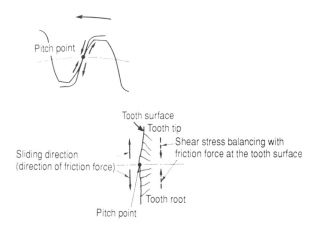

Figure 4.171 Plastic flow of tooth surface layer and direction of small cracks in the vicinity of pitch point

Table 4.36 *Mechanical properties**

	$\sigma_{0.2}$ (kgf/mm^2)	σ_B (kgf/mm^2)	El (%)	ϕ (%)	H_B	Impact value 2 mm U-notch (kgf m/cm^2)
Specification	≥ 55	≥ 80	≥ 12	≥ 45	235–388	≥ 5
Specimen	88	100	15	51	321	15

* Heat treatment: quenched after heating at 830°C for 4 hours and tempered at 180°C for 6 hours; carburized-hardened areas are excluded.

$\sigma_{0.2}$, proof stress; σ_B, tensile strength; El, elongation; ϕ, reduction in area; H_B, Brinell hardness number; RT, room temperature.

Table 4.37   *Study of design values*

Item	Unit	Design value	Allowable value
Peripheral speed	(m/s)	73.7	85 (MS company)
Bending stress of tooth	(kgf/mm²)	8.3	40 (MAAG)*
Surface pressure of tooth	(kgf/mm²)	71.6	145 (MAAG)
Resistance to pitting	(kgf/mm²)	89.8	145 (MAAG)

*MAAG: grinding company of gear in Switzerland.

In the case of the gear, wear and spalling were produced. In many cases, these phenomena are related to the hardness of the tooth surface. Accordingly, the carburizing and quenching conditions were investigated. Figure 4.172 shows the position where the hardness was measured, while Figure 4.173 shows the results of the hardness test. From the hardness distribution, it is apparent that the depth of layer with hardness $H_v \geq 500$ is about 0.5–1.3 mm, which is considerably smaller than the target depth of the effective hardened layer ($H_v \geq 500$, depth: 2 mm). The

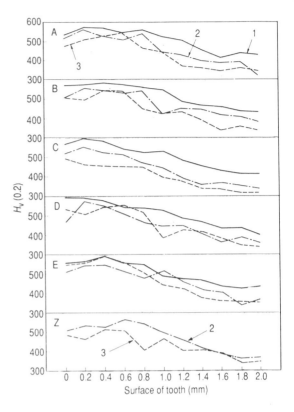

Figure 4.173   Hardness distribution at tooth surface

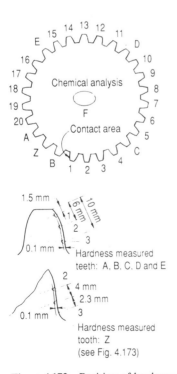

Figure 4.172   Position of hardness measurement

depth of the hardened layer is particularly small between the pitch point and the root and bottom. According to the results of analysis of carbon by X-ray microanalyser, both the amount and depth of carburization are small. The carburized condition at the root and bottom varies widely.

According to the results of observation of the fracture surface by SEM, the fracture surface is stepped in the range 1–2 mm from the tooth surface and a trace of plastic deformation is clearly seen. From these conditions, it is considered that the tooth surface received excessively large loads.

On the basis of the results of the investigation described above, it is considered that a change was caused in the contact condition due to abnormal wear at the root, and the pitch point was shifted in the direction of the tip of the tooth, causing localized loading at the tip and increasing the stress. The abnormal wear is attributable to insufficient lubrication, contact between the fracture surfaces and the application of too large a load for the material strength. In the present case, it is highly probable that failure is attributable to the latter cause.

(c) Cause and countermeasures

The failure of the gear is attributable to the cracking and spalling caused by excessive stress induced by wear and resultant localized loading because of low hardness of an insufficiently carburized tooth surface. To prevent this type of failure, the carburizing method should be improved so that the teeth can be uniformly carburized down to the bottom (target carbon content: 0.9%, target depth of carburization: 2 mm), and the tooth profile should be improved (tip and root of the pinion teeth). Moreover, lubrication should be controlled so that optimum lubricant temperature, optimum lubricant quantity and optimum lubricant spraying are ensured.

4.9.5 Case 4: pitting, dents and cracking in the tooth surface due to overloading and corrosion

(a) Outline of failure

Failure came about in the connecting gear between a steam turbine and an air turbo compressor. The failure is attributable to pitting, wear and dent-like indentations (about 14 mm long, 3 mm wide and 0.2–0.3 mm deep from the top end of the tooth to the fillet face) in the top face of the tooth, and cracks formed in many teeth in a direction vertical to the contact face (Figures 4.174–4.176). The service conditions of the gear are described below.

1. Number of revolutions: 11 000 rev/min
2. Air flow rate: 10 000 Nm³/h
3. Pressure: 6.53 atmospheric pressure gauge
4. Lubrication: 1.5 atg, 3 mm ϕ nozzle: about 4 l/min on one side, turbine oil No. 90, oil temperature: 55–60°C
5. Gear material: SCM445. It is suspected that the exhaust gas (hydrocyanic acid gas, 65°C) was mixed into the oil during quenching and tempering.

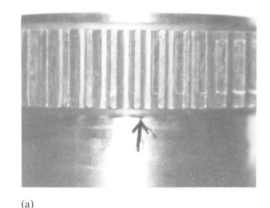

(a)

(b)

Figure 4.175 Outer appearance of broken portion (indicated by the arrow mark): (a) specimen 1; (b) specimen 2

(a)

(b)

Figure 4.174 Outer appearance of gear: (a) specimen 1; (b) specimen 2

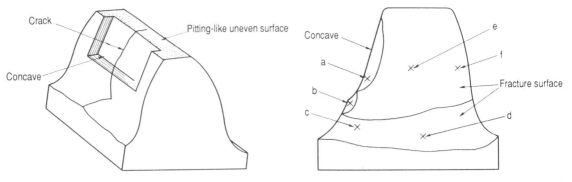

Figure 4.176   Outline of failure (specimen 1)

Figure 4.177   Observed positions of broken teeth (see Figure 4.178)

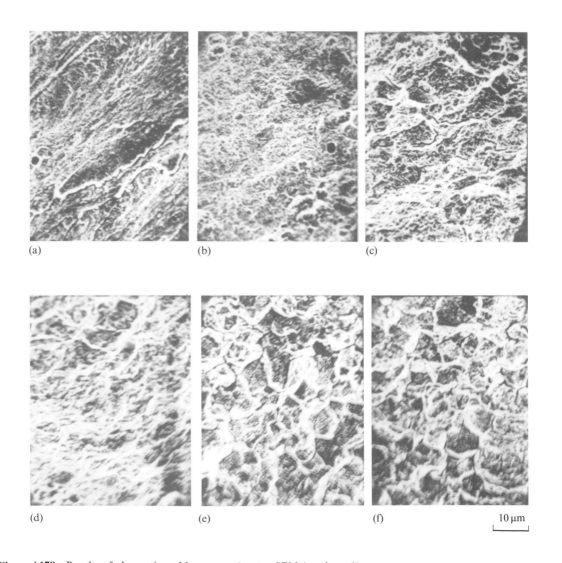

Figure 4.178   Results of observation of fracture surface by SEM (specimen 1)

(b) Results of investigation and discussion

The fracture surface was observed by SEM. The position where the fracture surface was observed is shown in Figure 4.177. The results of observation are shown in Figure 4.178. A trace of excessive plastic deformation is observed in the dent (Figure 4.178(a) and (b)). In the surface which was forcibly fractured by bending from the crack, a pattern of plastic deformation is observed, but the intergranular fracture surface accompanying secondary cracking predominates (Figure 4.178(c)–(f)). These results indicate that the teeth surface yielded at the dent due to excessive overloading, and the propagation of cracks is attributable to the environment. In the pitted part at the top face of the tooth, general corrosion and corrosion pits are observed.

The distribution of hardness in the tooth is shown in Figure 4.179. The whole tooth is hardened and the structure is a uniform tempered martensitic structure. No abnormalities are detected.

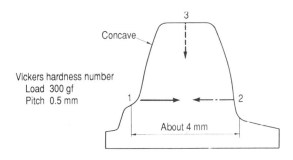

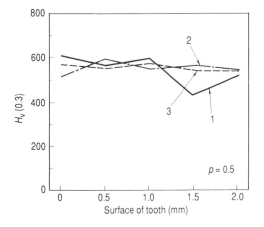

Figure 4.179   Hardness distribution of gear

On the basis of the results described above, it is apparent that failure is attributable to dents, pitting and cracking. Dents are attributable to load, while pitting and cracking corrosion are attributed to the environment. That is, dents are caused by overloading due to misalignment, while pitting and cracking are caused by corrosion and fracture in a corrosive environment. The hydrocyanic acid gas (HCN) contained in the exhaust gas is dissolved in the oil, forming H^+ and CN^- ions. H^+ causes hydrogen embrittlement in ferroalloys. Its sensitivity is particularly high in the case of high-strength materials. CN^- promotes corrosion through electrochemical and chemical reactions at the time of formation of characteristic complex compounds. It is considered that these phenomena are substantiated by the formation of pitting and an intergranular fracture surface.

(c) Cause and countermeasures

On the basis of the results of the investigation and discussion described above, it is considered that dents in the contact face of the tooth are attributable to excessive contact surface pressure due to misalignment caused during operation. It is estimated that pitting and cracking are caused by corrosion and hydrogen embrittlement attributable to the mixing of hydrocyanic acid gas (HCN) contained in the exhaust gas during lubrication.

To prevent this type of failure, the design and gear material should be changed. For the latter, it is desirable to perform nitriding or carburization to improve the hardness of the tooth surface. Nitriding or carburization reduces the wear and deformation of the surface and prevents hydrogen embrittlement by residual compressive stress.

4.9.6 Case 5: failure due to stress corrosion and corrosion fatigue

(a) Outline of failure

Five out of 24 teeth of the spiral gear of the driving shaft of a mixer were broken from the root fillet after about 4.5 months of service. Moreover, many cracks were detected at the same position in the spiral gear of the transmission shaft (see Figure 4.180).

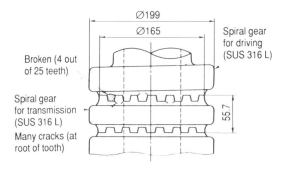

Figure 4.180   Schematic illustration and position of fracture of spiral gear

The service conditions of the gear are as follows.

1. Frequency: 30 Hz
2. Amplitude angle of driving shaft: $\pm 0.5°$
3. Environment: temperature 95°C; the liquid to be mixed is a corrosive material
4. Gear material: SUS316L for gears of driving and transmission shafts

(b) Results of investigation and discussion

The fracture surfaces of the gears of the driving shaft (fractured) and transmission shaft (cracked) were observed by SEM. The cracks caused at the root fillet are shown in Figure 4.181. Typical fracture surface patterns are shown in Figures 4.182 and 4.183. The cracks propagated in a direction 45° from the root fillet and a beach mark is observed in the fracture surface. This pattern is also observed in the gear of the driving shaft whose teeth were broken. Microscopic observation of the fracture surface revealed that cracks initiated at plural points, and a slip and intergranular fracture surface (Figure 4.182(b) and a slip and striation pattern (Figure 4.183(a)) are observed in the vicinity of the crack initiation points. Accordingly, the fracture surface is considered to be a fatigue fracture surface. In the crack propagation range, a cleavage fracture surface like a faucet pattern (flow pattern) is observed (Figure 4.182(c) and (d) and Figure 4.183(d)), and a mixture of striation and rub marks (a secondary fracture surface formed by friction between the fracture surfaces) is seen (Figure 4.182(e) and (f) and Figure 4.183(b) and (c)).

On the basis of the results of observation described above, it is considered that the causes of gear failure are fatigue due to repeated stress and stress corrosion cracking in a corrosive environment. Fracture caused by repeated application of stress in an environment where stress corrosion cracking is likely to be caused (stress corrosion cracking of austenitic stainless steel by Cl^- contained in the liquid to be mixed) is dependent largely upon the stressing condition. In the case of this gear, the applied stresses include a pressing force by the driving shaft, bending stress induced during operation, impeller vibration, etc. The stresses concentrate at the root fillet, producing a mixture of stress corrosion cracking and fatigue. This type of fracture surface is formed by changes in the stress wave form [5].

Accordingly, assuming that the stress applied to the gear is a trapezoidal stress wave (primary stress) on which the secondary sinusoidal wave stress (repeated stress) is superposed, it is estimated that a stress corrosion cracked surface is formed when the former stress predominates, but that a fatigue fracture surface is formed when the latter stress predominates.

(c) Cause and countermeasures

On the basis of the results described above, it is considered that failure of the gear is attributable to corrosion fatigue caused under stress corrosion cracking conditions. To prevent this type of failure, the gear material should be changed to martensitic stainless steel.

The service conditions of gears have become increasingly severe to meet the requirements for higher speeds and higher loads. Moreover, gears are used in a variety of environments. Since many factors are involved, it is important to clarify which is the main factor and which is the secondary factor of failure.

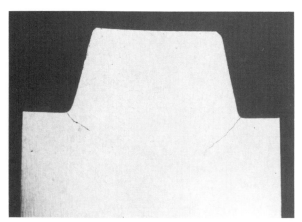

1 mm

Figure 4.181   Crack initiation at root of tooth (driving gear)

References

1. Nagao, Y. (1984) *Maintenance*, October issue, p. 49, Tokyo
2. Nagao, Y. (1984) *Maintenance*, November issue, p. 49, Tokyo
3. Subcommittee on Investigation of Causes of Gear Failures and Countermeasures (1975) *Report of Results of the Investigation*, p. 162, Tokyo
4. Aida, T. (ed.) (1974) *Design of Cylindrical Gears – Design and Fabrication of Gears*, Vol. 1, Ohkawa Shuppan, Tokyo, p. 123
5. Komai, K. (1981) *Corrosion Fatigue – Fractography and Its Application*, Nikkan Kogyo Newspaper Co. Ltd, Tokyo, p. 201

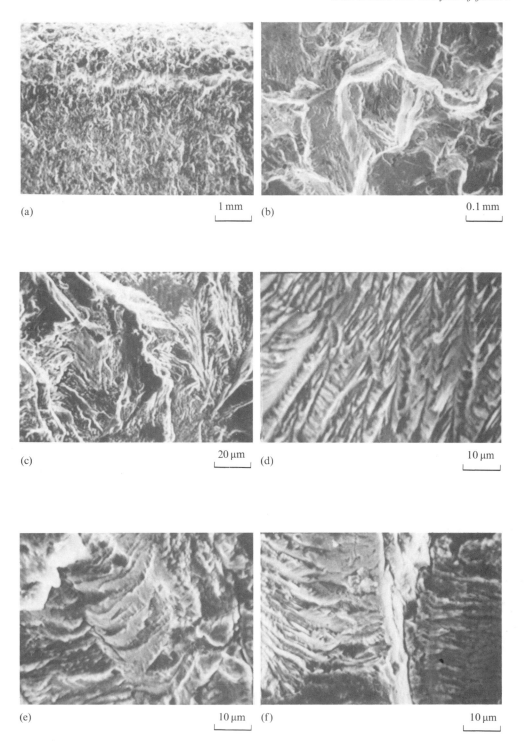

Figure 4.182   Results of observation by SEM of fracture surface of broken driving gear: (a) crack initiation area; (b) magnified (a); (c) crack growth area (intermediate stage); (d) crack growth area (intermediate stage); (e) crack growth area (final stage); (f) crack growth area (final stage)

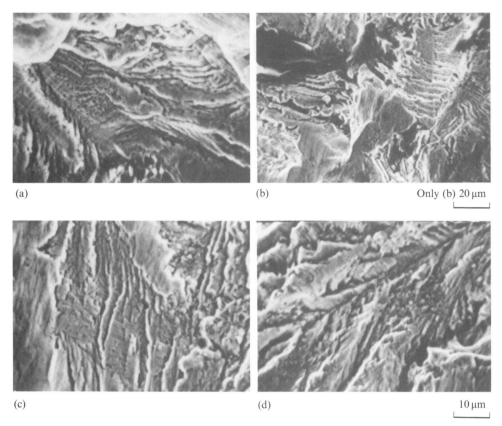

(a)

(b)

Only (b) 20 μm

(c)

(d)

10 μm

Figure 4.183   Results of observation by SEM of fracture surface of broken (transmission) gear: (a) crack initiation area; (b) crack growth area; (c) crack growth area (intermediate stage); (d) crack growth area (final stage)

4.10 Failure of piping valves [1,2]

At noon on a certain clear Sunday in the winter, a branch valve installed at a height of about 10 m from the ground suddenly fell. As the branch pipe had been removed before the occurrence of this failure, the valve was kept closed by attaching a blind patch to one side. From the appearance of the valve, it was considered that the material of the valve was a casting (graphite flake cast iron). Although macroscopic inspection revealed that the fracture surface is a brittle fracture surface, the failure initiation point could not be located.

At the present time, graphite flake castings are rarely used in main structural members. Because of their low cost and excellent castability [3], however, they are widely used in general members, such as piping valves, casting moulds and guide shoes. Failure of this type of member occurs frequently. Some work has been done on the failure of cast iron [4–12] but the majority of the work is directed towards spherical graphite cast iron

[4–10]. For graphite flake cast iron, however, little has been clarified, except that the fracture is spread through graphite flakes [4–6,10–12], because fracture surface analysis of cast iron is very difficult. There is no work directed towards quantitative analysis of fatigue failure and static failure.

With the opportunity presented by the failure of this valve, the failure of graphite flake cast iron was investigated from a fractographic standpoint with a view to reflecting the results of the investigation in the future analysis of failures. This section describes the results of the investigation.

4.10.1 Investigation of cause of failure

(a) Outline of failure
A diagram of the piping in which the valve was broken is shown in Figure 4.184. The broken valve is the branch valve installed between the large-capacity tank and the small-capacity tank. The failure of the valve is shown in Figure 4.185. The failure occurred in the centre of the casing. The casing was divided into three parts and fell.

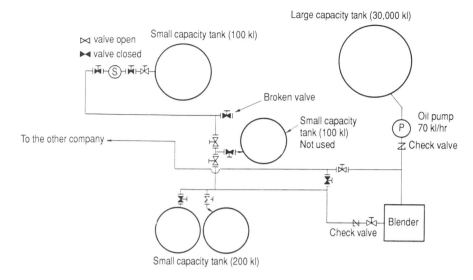

Figure 4.184   Diagram of piping system

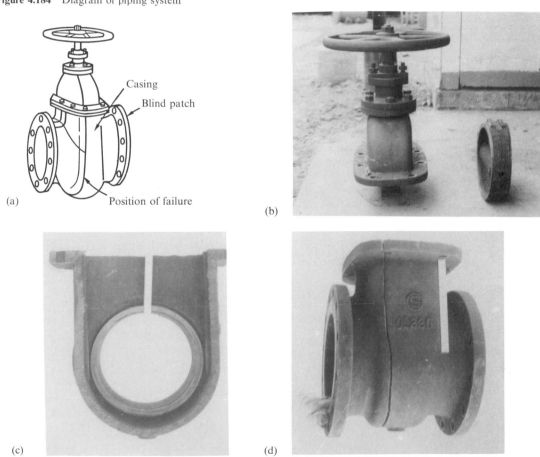

Figure 4.185   Outer appearance of broken branch valve: (a) sketch; (b) upper portion of broken valve; (c) fracture surface of valve; (d) lower portion of broken valve

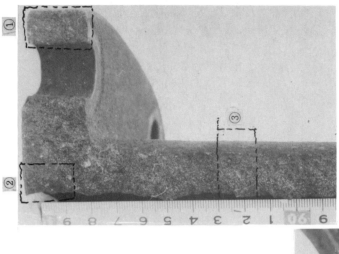

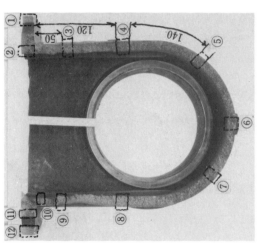

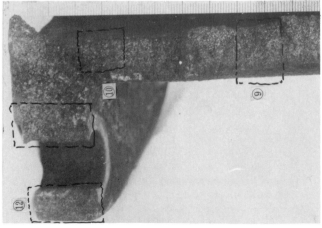

Figure 4.186 Macroscopic fracture surface and microscopically observed position

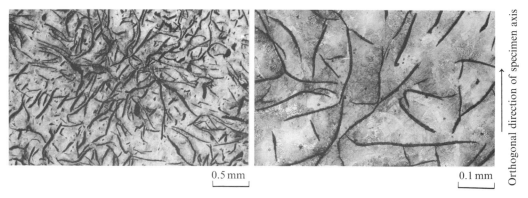

Figure 4.187 Microstructure of test specimen (at mid-thickness)

(b) Items investigated

1. Macroscopic observation
2. Chemical analysis
3. Mechanical properties
4. Microscopic observation (optical micrograph, SEM)
5. Hardness distribution

(c) Results of the investigation

(i) Macroscopic observation

Figure 4.186 shows the macroscopic fracture surface of the broken valve and the position of microscopic observation. Both macro- and microscopically, a large proportion of the fracture surface is a brittle fracture surface. The failure initiation point is not clear.

(ii) Chemical composition

The chemical composition and mechanical properties of the broken valve are shown in Table 4.38. The valve is made of graphite flake cast iron equivalent to FC10. The optical micrograph is shown in Figure 4.187. The structure is a typical graphite cast iron, containing a small quantity of bulk graphite. A large proportion of the matrix is pearlite with a small quantity of ferrite. Little difference if noted in the structure in the through-thickness direction, showing that the structure is comparatively homogeneous.

Table 4.38 *Chemical composition and mechanical properties of broken valve*

Chemical composition (wt %)					Mechanical properties	
C	Si	Mn	P	S	σ_B (kgf/mm^2)	El (%)
3.72	1.70	0.42	0.210	0.107	12.1	2.5

σ_B, tensile strength; El, elongation.

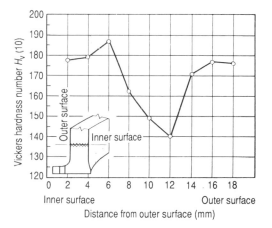

Figure 4.188 Hardness distribution in the vicinity of fracture surface

The hardness distribution in the vicinity of the fracture surface is shown in Figure 4.188. The hardness H_v is about 180 at the surface but is slightly lower in the centre.

The results of observation by SEM are described in Section 4.10.3 together with the results of the test. The cause of the valve failure could not be established from the results of the investigation described above. Accordingly, the following test and study were conducted.

4.10.2 Test method

(a) Material used

For this test, graphite flake cast iron FC10 (the sound part of the broken valve, 19 mm in thickness) was used. The test specimens were taken from about 10 mm in the centre. The chemical composition and mechanical properties are shown in Table 4.38 (see Figure 4.185).

(b) Profile of specimen

The profile of the specimen is shown in Figure 4.189. Figure 4.189(a) shows the specimens for the pulsating plane bending fatigue test (notched and plain specimens), while Figure 4.189(b) and (c) shows the specimens for the static bending test and the pulsating tensile fatigue test, respectively.

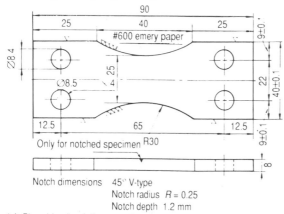

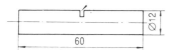

(a) Plane blending fatigue specimen (single side-notched or plain)

(b) Static bending specimen (single-side notched)

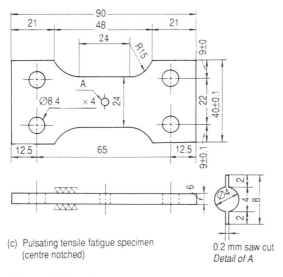

(c) Pulsating tensile fatigue specimen (centre notched)

0.2 mm saw cut
Detail of A

Figure 4.189 Dimensions of specimens and detail of notched area: (a) plane bending fatigue specimen (single side notched or plain); (b) static bending specimen (single side notched); (c) pulsating tensile fatigue specimen (centre notched)

(c) Test method

(i) Observation of the fracture surface

A pulsating plane bending fatigue test was conducted with a Schenck-type torsional fatigue tester (Figure 4.189(a)). The *S–N* curve was obtained for the plain specimen. When the fatigue crack length reached about 3.0 mm under repeated applications of a (net) stress $\sigma_n = 10\,\text{kgf/mm}^2$, which is higher by 1 kgf/mm² than the fatigue limit, the test was stopped and the specimen was fractured statically by bending. For the notched specimen, the stress ($\sigma_n = 10.5\,\text{kgf/mm}^2$) was repeatedly applied in such a way that a tensile stress was applied to the notched side. When the crack length from the bottom of the notch reached about 4.3 mm, the test was stopped and the specimen was fractured statically by bending.

For the static fracture test, an impact bending test was conducted by hammering one end of a round bar specimen, with the other end fixed in a vice (Figure 4.189(b)).

The fracture surfaces of the specimens were observed by SEM.

(ii) Measurement of the fracture surface ratio

Utilizing the photographs taken for the observation of the fracture surface in (i) above, the ratios of the part cracked by graphite and the uncracked part in the fatigue fracture surface and static fracture surface were determined. If the measuring range is narrow, exact values cannot be obtained. Accordingly, the measurement was made in an area of about $1.0 \times 1.6\,\text{mm}^2$ by joining the photographs taken at $200\times$. The photographs were joined, taking care not to overlap the printing papers, and the specific fracture surface was cut and divided with scissors. After measuring the weight of each piece of paper by a precision balance, the fracture surface ratio was calculated.

(iii) Observation of crack propagation route

A tensile pulsating fatigue test was conducted with an electro-hydraulic servo-type tension and compression fatigue tester (capacity: 10 tf) (Figure 4.190(c)). The

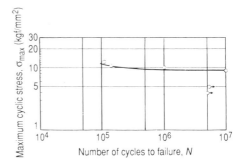

Figure 4.190 *S–N* curve for plain specimen by pulsating plane bending

stress ratio R (=minimum stress/maximum stress) was taken as 0.1 and the maximum stress (gross stress) $\sigma_g = 4$ kgf/mm^2 was repeatedly applied up to 50 000 cycles, while observing the crack propagating condition. After that, the stress was increased to 5 kgf/mm^2 and stressing was repeated up to 60 400 cycles in cumulative total. The stress was then further increased to 6 kgf/mm^2 and stressing was repeated up to 70 100 cycles. For observation of the crack propagation route, the tested parts were mirror-finished and the crack initiating and propagating conditions were examined by comparing the photographs taken before and after the test.

A static test was conducted with the same tester (capacity: 10 tf) until tensile failure occurred. The crack propagation route was observed by examining the photographs taken before and after the test (Figure 4.190(c)).

4.10.3 Results of the test and discussion

The testing conditions and the results of observation are summarized in Table 4.39.

(a) Fatigue strength in pulsating plane bending

The S–N curve of plain specimens obtained by the pulsating plane bending fatigue test is shown in Figure 4.190. Cast iron may fail at a number of cycles in excess of 10^7 cycles [6], but the maximum stress of 9.0 kgf/mm^2 below which the specimen does not fail at 10^7 cycles is taken as the fatigue limit. The ratio of this

fatigue limit to the tensile strength is $9.0/12.1 = 0.74$, which is slightly higher but is considered to be nearly reasonable. As the notch factor β of graphite flake cast iron is nearly 1 [6,13], the fatigue limit of the notched specimen is considered to be nearly equal. Accordingly, $\sigma_n = 10.5$ kgf/mm^2 was selected as the fatigue test stress for notched specimens.

(b) Observation of the fracture surface

Figures 4.191–4.193 show the fracture surface at a comparatively low magnification. Figure 4.191 is an example of a fatigue fracture surface, while Figure 4.192 shows an example of a surface which was statically fractured after a fatigue test. Figure 4.193 shows an example of a static fracture surface. In all the fracture surfaces, a smooth graphite fracture surface occupies quite a large area.

(i) Graphite surface

As is apparent from the results of observation of the crack propagation route described later, two types of crack propagate through graphite:

1. Cracks developed in the graphite, and
2. Cracks produced at the boundary between the graphite and the pearlite block.

If all fractures are to occur at the boundary between the graphite and the pearlite block, the area ratio of this fracture in the respective fracture surfaces will become equal. In practice, however, the area occupied by graphite is greater. This indicates that cracks are

Table 4.39   *Testing conditions and summary of observation*

Object	Type of specimen	Testing conditions (pulsating stress) (kgf/mm^2)	a (mm)	Summary of observation	
				Example	
Fractography, percentage of fracture surface	Single side notched	Pulsating plane bending fatigue	$\simeq 0$	Figure 4.187 (microstructure)	
		Figure 4.190(a),	1.1	Figure 4.191	
		$\sigma_n = 10.5$	2.0	Figure 4.194	Figure 4.196
		Static fracture after pulsating plane bending fatigue test $\sigma_n = 10.5$ Figure 4.190(a)	5.6	Figure 4.192	
			6.0	–	
		Static bending	$\simeq 0$	Figure 4.193	
			1.0	–	
		Figure 4.188(b)	5.9	Figure 4.195	Figure 4.197
S–N curves (fractography)	Plain	Pulsating plane bending fatigue		Figure 4.198	
	Figure 4.190(a)	$\sigma_n = 10.0$			
Crack propagation path	Centre notched Figure 4.190(c)	Pulsating tensile fatigue $\sigma_g = 4.0$–6.0 Static fracture (tension)	$\simeq 0$ $\simeq 0$	Figure 4.199 Figure 4.201	Figure 4.200 Figure 4.202

σ_g, gross stress; σ_n, net stress; l_f, fatigue crack length (mm)
a, observation location from notch bottom (mm), i.e. $a \simeq 0$ means at the vicinity of the notch bottom.

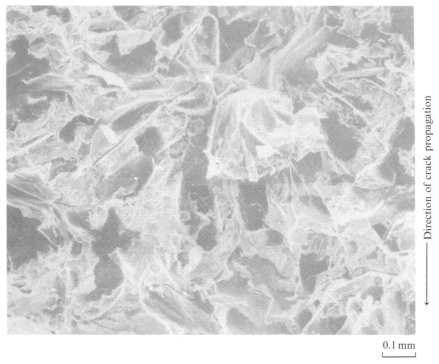

0.1 mm

Figure 4.191   Fatigue fracture surface (single side notched, $a = 2.0$ mm, $l_f = 4.3$ mm $\sigma_n = 10.5$ kgf/mm²)

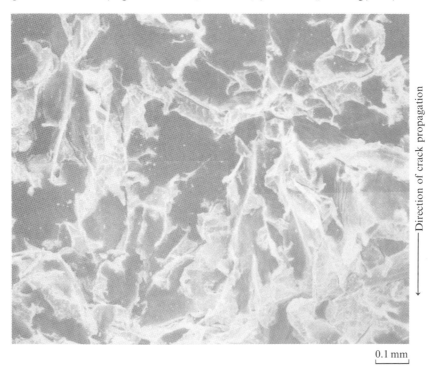

0.1 mm

Figure 4.192   Static fracture after fatigue test (single side notched, $a = 5.6$ mm, $\sigma_n = 10.5$ kgf/mm², $l_f = 4.3$ mm; see Table 4.39)

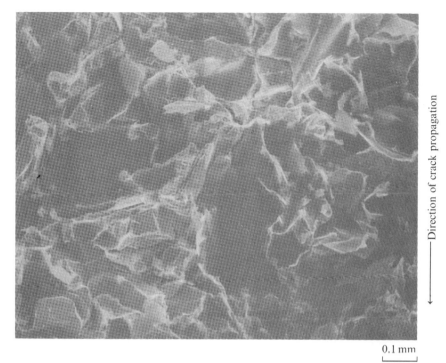

Direction of crack propagation

0.1 mm

Figure 4.193 Static fracture surface (single side notched; in the vicinity of notch, $a \simeq 0$)

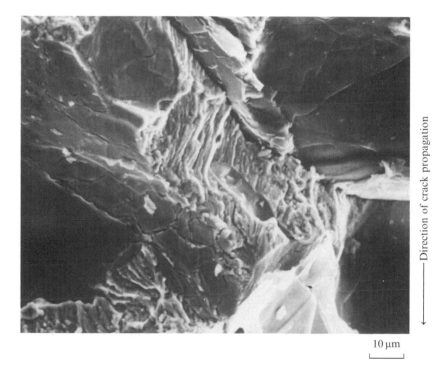

Direction of crack propagation

10 μm

Figure 4.194 Fatigue fracture surface (single side notched; in the vicinity of notch, $a \simeq 0$)

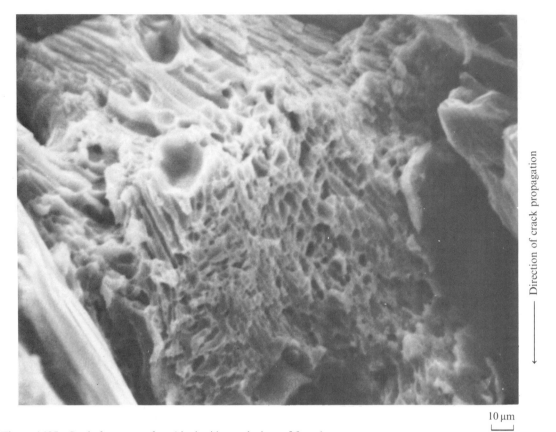

Direction of crack propagation

10 μm

Figure 4.195   Static fracture surface (single side notched, $a = 5.9$ mm)

also developed in the graphite. That is, this indicates that there are two types of crack which propagate through graphite.

(ii) Cracking of the pearlite block
A cementite layer in pearlite is observed in the fatigue fracture surface shown in Figure 4.194 and the static fracture surface shown in Figure 4.195. The lamellar part which is seen in the centre of Figure 4.195 is considered to be a typical fatigue surface [14] (Y. Fujimura, H. Ishii and M. Kawarazaki, 1976, private communication). A dimple fracture surface is seen on the right of the centre. As is seen from Figure 4.197, the dimple fracture surface occupies a large proportion of the part where cracks are caused from pearlite in the static fracture surface. Accordingly, it is considered that the specimen was fractured statically.

On the basis of the above description, the fracture surface can be judged as a fatigue fracture surface if the stratiformed part shown in Figure 4.194 occupies a large proportion of the part where cracks are caused from pearlite, but as a static fracture surface if the dimple fracture surface occupies a large proportion.

Figure 4.196 shows the quasi-cleavage of pearlite in

a fatigue fracture surface. Figure 4.197 shows the quasi-cleavage of pearlite in a static fracture surface. A lamellar pearlite fracture surface and a dimple fracture surface described previously are observed.

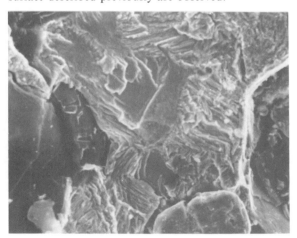

Figure 4.196   Fatigue fracture surface (single side notched, $a = 1.1$ mm, $l_f = 4.3$ mm)

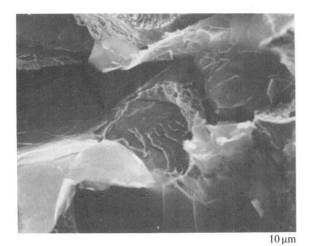

10 µm

Figure 4.197   Static fracture surface (single side notched, $a = 5.9$ mm)

A dimple fracture surface and quasi-cleavage occupy a large proportion of the fatigue fracture surface. This indicates that the fatigue cracks propagate nearly statically from one graphite area to an adjacent graphite area in one cycle of stressing on their way through the graphite, or fatigue cracks propagating slowly from adjacent graphite areas are momentarily connected with each other in one cycle.

(c) Type of failure and pearlite fracture surface ratio

The graphite fracture surface ratio and pearlite fracture surface ratio are defined as follows. That is, the ratio of the fracture surface in the part where cracks are produced through the graphite is called the graphite fracture surface ratio, while the ratio of the fracture surface in the part where cracks are produced in the pearlite block or at the boundary between pearlite blocks is called the pearlite fracture surface ratio. The graphite fracture surface ratio is the sum of the fracture surface ratios in the part which is cracked in the graphite and in the part which is cracked at the boundary between the graphite and the pearlite block. Addition of the graphite fracture surface ratio and the pearlite fracture surface ratio makes 100%.

Figures 4.191–4.193 were used for measurement of the fracture surface ratio. The results of measurement are summarized in Table 4.40. As shown in Table 4.40, the graphite fracture surface ratio is 60–90%. Since the ratio determined from observation of the structure of cast iron is only several percent [13], the ratio measured by us is far too high, indicating that failure proceeds through the graphite. The pearlite fracture surface ratio is 34–41% in the fatigue fracture surface but as low as 11–14% in the static fracture surface. The latter ratio is 1/3–1/4 of the former ratio. That is, a clear difference is observed between the two ratios. This means that fatigue failure differs in mechanism from static failure and this difference is a key point for identification of the type of failure.

In the case of fatigue failure, the crack initiates at the end of the graphite and propagates through the matrix under the localized dynamic condition prevailing at the end of the graphite or in the vicinity of the crack tip. After propagation for some distance, the crack connects with the crack propagating from the end of the adjacent graphite area or with the adjacent graphite area. In this case, therefore, the crack does not

Table 4.40   *Percentages of graphite fracture surface and pearlite fracture surface*

Type of specimen	a (mm)	Fatigue fracture surface		Static fracture surface	
		Percentage of graphite fracture surface	*Percentage of pearlite fracture surface*	*Percentage of graphite fracture surface*	*Percentage of pearlite fracture surface*
Single side notched	0	62.8	37.2		
Pulsating plane	2.0	66.2	33.8		
bending ($l_f = 4.3$)	4.0	65.9	34.1		
	5.6			81.2	18.8
	6.4			78.0	22.0
Plain	0	59.2	40.8		
Pulsating plane	2.0	66.5	33.5		
bending ($l_f = 3.0$)	4.0			87.1	12.9
	6.0			87.7	12.3
Static bending	0			89.5	10.5
	7.4			86.4	13.5
Fracture surface				91.0	9.0
of valve				90.4	9.6

l_f, fatigue crack length (mm), a, observation location from the notch bottom (mm).

necessarily propagate through the shortest distance between graphite areas in three dimensions. In the case of static failure, however, the connection of graphite areas by a crack is governed by the dynamic condition prevailing in a more macroscopic range, including both graphite areas. Accordingly, the crack propagates through the shortest route between graphite areas. This is the reason why the pearlite fracture surface ratio for static fracture is smaller than that for fatigue fracture. As shown in Table 4.40, the pearlite fracture surface ratio for static fracture after the fatigue test is about 20%, which is greater than that for simple static fracture (by some 10%). This means that a fatigue crack initiates at graphite and propagates with a considerable proportion of fatigue fracture surface mixed therein.

Figure 4.198 shows the difference in the failure mechanism described above. It will be well understood from Figure 4.198 that the crack propagation route in fatigue fracture differs from that in static fracture and the pearlite fracture ratio measured in the fatigue fracture surface is smaller than that measured in the static fracture surface.

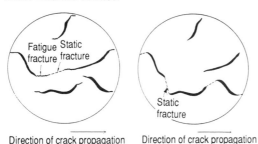

(a) Crack propagation mechanism in fatigue fracture

(b) Crack propagation mechanism in static fracture

Figure 4.198   Schematic illustration showing the difference between fatigue fracture and static fracture. Crack propagation mechanism in (a) fatigue fracture and (b) static fracture

(d) Crack propagation route

The results described in (c) above are substantiated below. Figures 4.199 and 4.200 show the crack propagating condition on one side of the specimen in the pulsating tensile fatigue test, while Figures 4.201 and 4.202 show the same condition in the static tensile test. In each photograph, a notch is shown at the upper left and the macroscopic direction of crack propagation is from left to right. Figures 4.199 and 4.201 show the crack propagation route, which was determined on the basis of Figures 4.200 and 4.202 respectively. Figures 4.199 and 4.201 clearly show that cracks propagate through the graphite. A comparison between fatigue fracture and static fracture shows that cracks in static fracture propagate through graphite by a shorter

route, i.e. by the shortest route in the matrix. As described previously, this phenomenon substantiates the fact that the pearlite fracture surface ratio for the static fracture surface is smaller.

In Figure 4.200 the initiation of many fine cracks at the graphite is observed in addition to the main crack which initiates at the notch bottom and propagates. During the application of a primary stress σ_{g1} of 4.0 kgf/mm^2 for $N_1 = 50\,000$ cycles, the crack stopped propagating. The crack did not propagate even when the stress was increased to 5.0 kgf/mm^2 and repeatedly applied for $N_2 = 10\,000$ cycles. Accordingly, the stress was further increased to 6.0 kgf/mm^2 and repeatedly applied for $N_3 = 10\,000$ cycles. Shown in Figure 4.199 is the condition when a stress of 6.0 kgf/mm^2 was applied for 10 000 cycles. After this stressing, the crack which had been stopped started to propagate again. Nisitani and Ohi [15] reported the initiation of a non-propagation crack at graphite in the notched and plain specimens of a similar cast iron.

(e) Comparison with actual failures

At the bottom of Table 4.40 are the graphite fracture surface ratio and the pearlite fracture surface ratio which were obtained from observation of the part of the fracture surface of the broken piping valve where failure is most likely to initiate. As these values are nearly equal to the values obtained from the surface which was fractured statically by bending in the laboratory, it may be said that the valve was statically broken due to increases in internal pressure. These values also substantiate the effectiveness of the analysis method employed by us.

4.10.4 *Summary and countermeasures*

As the cause of failure of the broken valve and the failure initiation point in the valve could not be established from macro- and microscopic observation, the fracture surfaces of the fatigue test specimen and static bending test specimen taken from the valve were analysed fractographically. As a result, it was found that the pearlite fracture surface ratio for the fatigue fracture surface is 34–41% and that for the static fracture surface is 11–14%. The pearlite fracture surface ratio determined by observation of the part of the fracture surface of the broken valve where failure is most likely to initiate was 9.0–9.6%. This ratio is nearly equal to that obtained after static bending fracture in the laboratory. It was therefore clear that the valve was statically broken as a result of increases in internal pressure.

As countermeasures:

1. Relief valves are installed at certain points of the piping so that the piping system is not closed
2. Pressure gauges are installed at several points in the piping system to control the pressure in the system.

Notch bottom

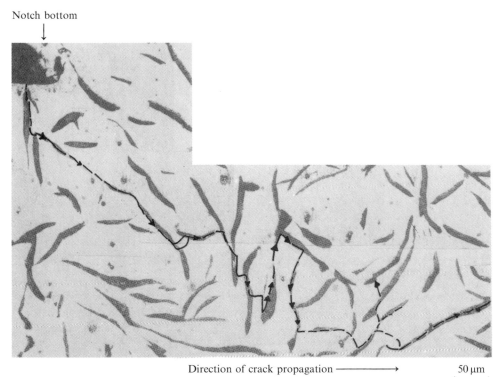

Direction of crack propagation ⟶   50 μm

Figure 4.199   Fatigue crack propagation route (centre notched, before test)

Notch bottom

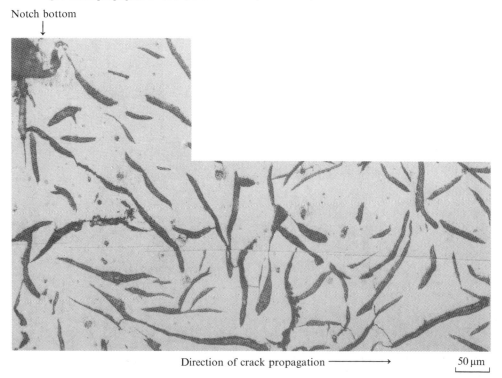

Direction of crack propagation ⟶   50 μm

Figure 4.200   Fatigue crack propagation route (centre notched; stress and number of cycles: $4.0 \times 50\,000 \rightarrow$ $5.0 \times 10\,400 \rightarrow 6.0$ kgf/mm$^2 \times 9700$ cycles; see Figure 4.199)

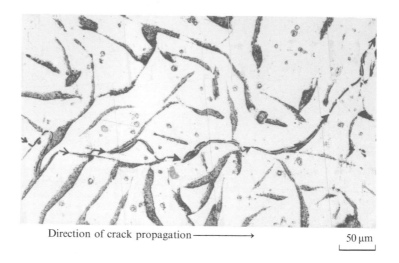

Direction of crack propagation ⟶   50 μm

Figure 4.201   Static crack propagation route (centre notched, in the vicinity of notch, before test)

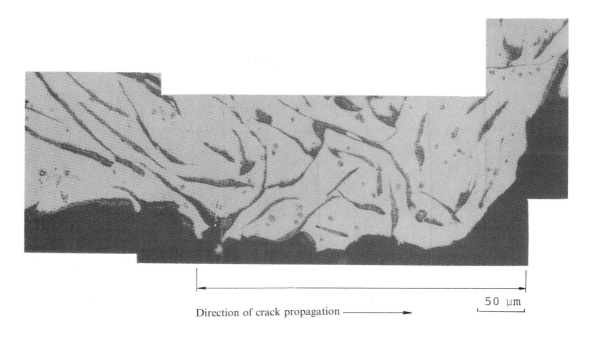

Direction of crack propagation ⟶   50 μm

Figure 4.202   Static crack propagation route (centre notched, in the vicinity of notch bottom, after test; see Figure 4.201)

References

1. Nishida, S. Urashima, C. and Masumoto, H. (1981) In *Proceedings of the 26th Japan National Symposium on Strength, Fracture and Fatigue*, Sendai, Japan, p. 91
2. Nishida, S., Urashima, C. and Yonekura, T. (1982) *Journal of the Society of Materials Science*, **31**, 922
3. Journal of the Society of Materials Science (ed.) (1972) *Materials for Machines and Their Test Method*, Kyoto, Japan, p. 207
4. Kobayashi, T. (1973) *Journal of the Iron and Steel Institute of Japan*, **59**, 1578, Tokyo

5. Igawa, K. and Tanaka, Y. (1974) *Journal of the Japan Society of Metals*, **13**, 665, Sendai, Japan

6. Tanaka, Y., Honma, T. and Igawa, K. (1975) *Castings*, **47**, 338, Tokyo

7. Sunada, H. Abstract of the report to the Fractography Subcommittee (81973), The Japan Society of Mechanical Engineers, Tokyo, p. 125

8. Sunada, H. (1977) *Journal of the Society of Materials Science*, **22**, 767, Kyoto, Japan

9. Ohi, T. and Nagao, Y. (1977) Preprints for 1st Fractography Symposium, 11, Journal of the Society of Materials Science, Kyoto

10. Arii, M. and Tajima, K. (1977) *Journal of the Society of Materials Science*, **26**, 767

11. Shiota, T. and Komatsu, S. (1977) *Castings*, **39**, 602, Tokyo

12. Shiota, T. and Komatsu, S. (1978) *Journal of the Society of Materials Science*, **27**, 291, Kyoto, Japan

13. Casting Association (1955) *Properties of Castings*, Coronasha, Tokyo, pp. 19 and 105

14. Nishida, S., Urashima, C. Sugino, K. and Masumoto, H. (1979) In *Proceedings of the International Conference on the Strength of Metals and Alloys*, **5**, 1255

15. Nisitani, H. and Ohi, H. (1979) *Transactions of the Japan Society of Mechanical Engineers*, **790 2**, 128

4.11 Failure of pipeline before service [1]

4.11.1 Introduction

With the increasingly severe conditions of development of energy resources, such as oil and natural gas, exploration is extending into remote areas and deep waters. With these changes in the conditions of exploration and development of energy resources, various new problems have arisen.

The problem discussed in this section is the water leakage which resulted at a water pressure far lower than the target pressure during a water pressure test of a pipeline to be installed in Alaska for the transportation of natural gas. When the leakage was detected, the people on site inferred that it was attributable to defects caused at the time of pipe manufacture. However, the quality inspection of the pipe after manufacture did not reveal any defects. Moreover, later inspection revealed some sign of damage on the outside surface of the pipe from which water leaked. It was therefore suspected that the leakage might be attributed to trouble arising after pipe manufacture.

The results of the investigation of the cause and analysis, and the mechanism by which cracks initiated and propagated in the pipe are described below, clarifying that the main cause of water leakage was failure of the pipe brought about during marine transportation.

4.11.2 Methods of investigation and test

(a) Investigation of points from which water leaked

Water leaked from five points on the entire line. For all these points, macroscopic investigation (appearance inspection, X-ray inspection, magnaflux test, S-print) and microscopic observation (structure of the cross section of the broken part, observation of the fracture surface by SEM, EPMA, etc.) were conducted.

(b) Laboratory fatigue test

Figure 4.203 shows the pipeline installed on site. Damage was detected at the points where water leaked, and this leakage was supposed to be attributable to fatigue according to the results of macroscopic observation. Accordingly, the loads applied to the points were calculated, and a completely reversed plane bending fatigue test and pulsating bending fatigue test of the actual pipe were conducted. The former test was conducted to obtain the fatigue strength of the pipe. For the test, test specimens were taken from the sound portion of the pipe from which water leaked. The completely reversed plane bending fatigue test was conducted on the basis of the results of analysis of a fracture surface which are described later (for example, see Figure 4.211). The test specimen and

Figure 4.203 Laying of pipeline on site

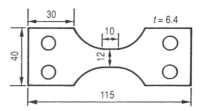

Figure 4.204 Dimensions of plain bending fatigue test specimen

testing conditions are shown in Figure 4.204 and Table 4.41 respectively. It was supposed that quite a large load was applied in a concentrated way to specific parts of the pipe during transportation. Accordingly, the latter test was conducted by the three-point bending method as shown in Figure 4.205. The conditions of the test are shown in Table 4.42. For the test, a pipe of the same size and same steel grade as that of the pipe from which water leaked (except E in Table 4.43) was used. The stresses induced at various parts of

Figure 4.205   Pulsating bending fatigue test of actual pipe

Table 4.41   *Conditions of reversed plane bending fatigue test*

	Test conditions	
Type of stress	Frequency	Stress amplitude (kgf/mm²)
Completely reversed	1500 cycles/min	(H-1) 35 (H-2) 30 (H-3) 28

1. Test specimen: cut out from the normal part of the received pipe D.
2. Dimensions of specimen: 12 mm width with skin (cut out from the flattened pipe with original thickness).
3. Testing machine: Schenk-type fatigue tester.

Table 4.42   *Example of conditions for pulsating bending fatigue of actual pipe*

Type of stress	Frequency (cycles/min)	Stress amplitude (kgf/mm²)
Constant stress amplitude	1300	30.0

1. Dimensions of pipe: $6\frac{5}{8}$ (outer diameter) × 0.375 (thickness) × 20 (length) (unit: inches).
2. Testing, machine: Schenck-type tensile and compressive fatigue tester

Table 4.43   Conditions of water leakage

Dia. and wall thickness	Testing pressure before shipment (psi)	Water pressure by leakage (psi)	Leaked portion of water
A φ6 ⁵/₈" × 0.375"	6630	2525	39' 3"|3" / 36'6"
B φ6⁵/₈" × 0.250"	4420	1065	37'9" 3"|2'6" / 40'8"
C φ6⁵/₈" × 0.250"	4420	1065	4'9"|4" 35'5" / 40'8"
D φ6⁵/₈ × 0.250"	4420	770	13'5" 6" 26'3" / 39'8"
E φ16" × 0.406"	2970	—	No data available

the pipe at the time of loading were measured by strain gauges. After the test, the specimens were observed macroscopically and microscopically and the results of the observation were compared with those of the actual pipe from which water leaked.

(c) Investigation of the pipe material
To grasp the basic characteristics of the pipe material by investigating its properties, a tensile test, hardness measurement, impact test and microscopic observation were conducted with a specimen taken from that portion of the pipe near the point of leakage which was estimated to be a sound portion from visual observation.

4.11.3 Results of the investigation and test

(a) Leakage of water
The leakage of water which was caused during the hydraulic pressure test of the pipeline on site is shown in Table 4.43. In all the pipes, water leaked at a pressure of 15–40% of the pressure used in the test conducted before shipment. Water leakage is not limited to the specific points in the pipe and to the specific sizes of pipe.

(b) Macroscopic observation of point of leakage
Figure 4.206 shows the appearance of a point from which water leaked. A local dent is observed and several sharp cracks extending from the dent in the longitudinal direction of the pipe are also observed. As such local dent and sharp cracks are detected at other

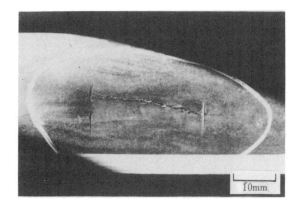

Figure 4.206 Outer appearance of portion of pipe B where leakage of water occurred

points from which water leaked, it is supposed that this damage was caused under similar conditions. A non-destructive test revealed that the dents and cracks had arisen only at the points from which water leaked.

(c) Observation of the cross section of the broken part

Figure 4.207 shows an example of the structure of the cross section of the broken part, including cracks. The cracks initiated at the outside and inside surfaces of the pipe and propagated nearly vertically to the surface towards the centre in the through-thickness direction. The cracks propagated in the ferrite grains and grain boundaries. Some cracks were branched on the way.

In the cracked surface, the existence of oxides is observed up to the tip of the crack, but decarburization is not seen around the cracks.

Figure 4.208 shows the outside surface of the pipe in the vicinity of the broken point. A systematic plastic strain which is considered to be caused by cold working is observed.

Figure 4.209 shows the metal flow (etched by Oberhoffer's solution [2]) caused by rolling at the time of pipe manufacture. It is apparent that the propagation of cracks is not related to the metal flow. The same tendency was observed at other points by microscopic observation. On the basis of the results described above, it is estimated that failure and cracks at these points were caused after pipe manufacture (heating→ rolling→ heat treatment→ finishing→ quality inspection). In other words, cracks initiated at the outside and inside surfaces and propagated under an external force which was applied during transportation and caused plastic deformation of the outside surface.

(d) Observation of the fracture surface

Since the fracture surface was so badly corroded that it could not be used for observation, rust was removed from the fracture surface by pickling. Figures 4.210 and 4.211 show examples of the results of observation of the fracture surface. Cracks initiated at the outside and inside surfaces of the pipe. In these examples, the final fracture position is nearly in the centre of the wall thickness. The fracture surface is comparatively smooth and the cracks join up with each other during

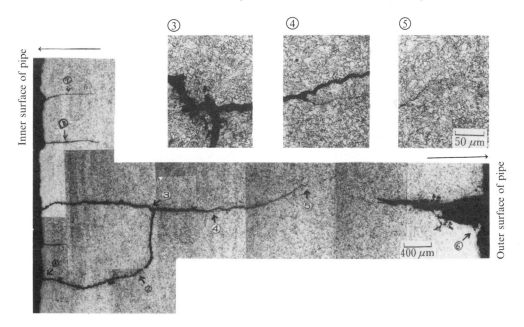

Figure 4.207 Microscopic structure (in vertical section) of broken portion of pipe B

Outer surface of pipe

50 μm

Figure 4.208   Outer surface of pipe B in the vicinity of broken portion

Outer surface of pipe

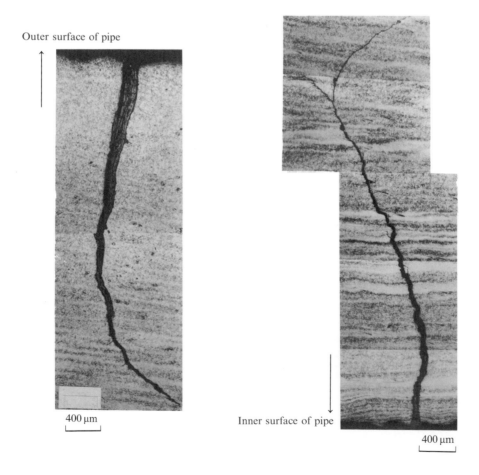

400 μm

Inner surface of pipe

400 μm

Figure 4.209   Metal flow structure by rolling in cross section of broken portion of pipe A

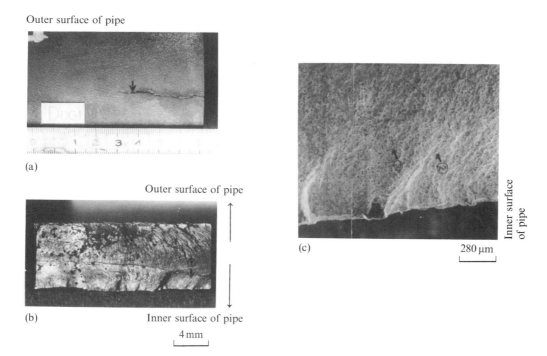

Figure 4.210   Outer appearance of fracture surface (pipe D): (a) outer surface of pipe D; (b) fracture surface; (c) enlarged view of (b)

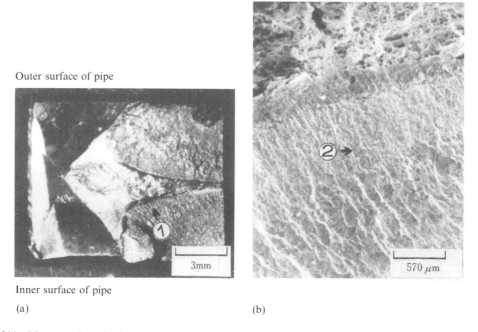

Figure 4.211   Macroscopic and microscopic fracture surface (pipe E): (a) outer appearance of fracture surface; (b) enlarged view of point 1

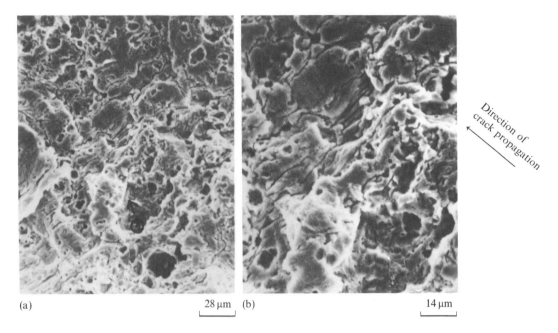

(a)   28 μm   (b)   14 μm

Figure 4.212   Results of observation of fracture surface by SEM (pipe E): (a) enlarged view of point 2 in Figure 4.211(b); (b) enlarged view of (a)

propagation. The fracture surface resembles the fatigue fracture surface which is produced by repeated load.

Part of the fracture surface shown in Figure 4.211 was observed by a microscope with higher magnifying power, the results of which are shown in Figure 4.212. Many corrosion pits are observed on the fracture surface but cleavage and intergranular fracture surfaces are not observed at all. Figure 4.212(b) is an enlarged view of Figure 4.212(a). A streak pattern which is perpendicular to the macroscopic direction of crack propagation is observed. The pattern is estimated to be striation which is peculiar to a fatigue fracture surface. If the pattern matches the type of macroscopic fracture surface shown in Figure 4.211, it may be said that the fracture at the point where water leaked is a fatigue fracture caused by repeated load. In the striation pattern, flat portions are observed here and there. It is considered that these flat portions were formed by partial crushing of the striation caused by the contact between the mating fracture surfaces after cracking. Accordingly, the streak pattern is judged to be striation due to fatigue fracture. According to the results of observation of other points by SEM, this streak pattern could not be detected at other points, probably because of severe corrosion of the fracture surfaces. However, as all macroscopic fracture surfaces are similar to those shown in Figure 4.210, it may safely be said that they are fatigue fracture surfaces.

According to the results of EPMA of the cross section and the fracture surface, including cracks, no inclusions or trace of inclusions could be detected. It can be concluded that the base material of the pipe is free from defects.

On the basis of the results described above, water leakage caused during the hydraulic pressure test which was undertaken before putting the pipeline into service is attributable to fatigue caused by repeated applications of load to the pipe during transportation. As the fracture surfaces were severely corroded, it may be said that failure was promoted by a corrosive environment.

4.11.4 Study of the cause of failure

As a result of observation of the points from which water leaked, it was found that the pipe failure is a fatigue fracture caused after shipment. Since the water leakage occurred before the pipes were put to use, it is inferred that the failure occurred during transportation from Japan to Alaska. However, there are some questionable points. The first question is whether there is any possibility that fatigue fracture of a sound pipe is caused by repeated application of loads developed by rolling of the ship during transportation, etc. The second question is why cracks initiate and propagate from both the outside and inside surfaces of the pipe when the pressing load is applied to the outside surface of the pipe due to contact of the pipe with other pipes or other materials. These points must be clarified.

To elucidate the cause of the failure, the following laboratory tests were conducted.

(a) Fatigue characteristics of the pipe

Since it was found that the failure of the pipe is a fatigue fracture, the fatigue characteristic of the base material of the pipe was investigated as one of the basic characteristics. The results of the investigation are shown in Figure 4.213. From Figure 4.213 the fatigue characteristic of the base material of the pipe is the same as that for HT60, having the same strength level. Accordingly, the fatigue characteristic of the pipe is satisfactory.

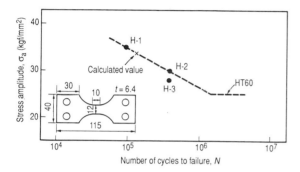

Figure 4.213 *S–N curve for sound portion of pipe D*

(b) Estimation of repeated stress induced during transportation

Separately from the laboratory test of the actual pipe, the stresses applied to the pipe were estimated by the method of fracture mechanics based on the results of observation of the fracture surface by SEM. That is, the surfaces with striation (Figures 4.211 and 4.212) were used for estimation.

The crack propagation in the fracture surfaces is shown schematically in Figure 4.214. The point A is the point where striation is observed, while the point B is the point for evaluation in calculations (both points are on the same ellipse). For calculation, it is assumed that the crack propagates in a semi-ellipsoidal manner and that a surface notch exists in the part where the bending moment is applied. For this solution, the equations developed by Raju and Newman [3] and

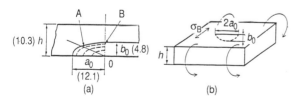

Figure 4.214 Model for crack propagation in pipe

Ishida and Noguchi [4] are applied. For our calculation, a strict solution by the latter equation is used. According to this method, the stress intensity factor K at the point B is given by

$$K = \sigma_B \sqrt{(\pi)} \cdot A \tag{4.46}$$

where the parameters are:

σ_B surface stress induced by bending
l depth of the crack
A $1.1359 - 0.3929\mu - 0.3440\mu^2 + 0.2613\mu^3$
  $+ \lambda(-1.5184 + 0.4178\mu + 0.7846\mu^2 - 0.6329\mu^3)$
  $+ \lambda^2(4.3721 - 13.9152\mu + 16.2550\mu^2 - 6.4894\mu^3)$
  $+ \lambda^3(-3.9502 + 12.5334\mu - 14.6137\mu^2$
  $+ 5.8110\mu^3)$
μ l/a ($2a$ is the width of the crack)
λ l/h (h is the wall thickness of the pipe)

The crack propagation rate is given by

$$dl/dN = C(\Delta K)^m \tag{4.47}$$

where N is the number of cycles to failure and C and m are constants for the material ($C = 1.0 \times 10^{-10}$, $m = 3$ (see Section 3.3.2).

However, the striation distance at the point A, S_A, is 1.0×10^{-3} mm, from Figure 4.210. Accordingly, $S_A = dl/dN$. Substituting S_A into equation (4.47) the K-value at the point A is calculated as shown below:

$$\Delta K_A = (1.0 \times 10^{-3} / 1.0 \times 10^{-10})^{1/3}$$
$$= 215.4 \text{ kgf/mm}^{3/2}$$

Assuming that the bending load is applied to the plate with a semi-ellipsoidal surface crack, the above K-value is converted into the K-value at the deepest point of the crack (point B in Figure 4.212) by using Smith's solution [5]:

$$\Delta K_B = 0.5 \times \Delta K_A = 107.7 \text{ kgf/mm}^2$$

From this value, the repeated stress σ_a ($= \sigma_B/2$) is obtained by the equation (4.47) for strict solution which was developed by Ishida *et al.* Then:

$$\sigma_a = \frac{\sigma_B}{2} = \frac{K}{2\sqrt{\pi l \cdot A}} = \frac{107.7}{2\sqrt{\pi} \times 4.8 \times 0.484}$$
$$= 28.6 \text{ kgf/mm}^2 \tag{4.48}$$

The number of cycles to failure is then estimated. From equation (4.47):

$$dN = \frac{dl}{C(\Delta K)^m}$$

Therefore

$$N = \int_{b_i}^{b_0} \frac{dl}{C(\Delta K)^m} = \int_{b_i}^{b_0} \frac{dl}{C(\sigma_B A)^m \pi^{m/2} l^{m/2}}$$

$$= \frac{1}{C(\sigma_B A)^m \pi^{m/2}} \frac{1}{(m/2) - 1} \left(\frac{1}{b_i^{m/2-1}} - \frac{1}{b_0^{m/2-1}} \right)$$

$$=\frac{1}{1.0 \times 10^{-10} \times (57.2 \times 0.484)^3 \times \pi^{3/2}}$$

$$\times \frac{1}{0.5} \left(\frac{1}{0.5^{1/2}} - \frac{1}{4.8^{1/2}} \right)$$

$$=16.2 \times 10^4 \tag{4.49}$$

where b_i is the depth of the virtual initial crack. Judging from the fact that the fracture surface is considerably corroded, it is reasonable to consider that the depth of the initial crack is about 0.5 mm. Since the pipe material is free from initial defects, however, the number of cycles to the formation of a 0.5 mm deep crack can be considered to be about 50% of the total number of cycles to failure [6]. Accordingly, the total number of stress cycles applied to the pipe can be obtained by doubling the value calculated by equation (4.49). That is, the number is estimated to be 32.4×10^4 cycles.

(c) A pulsating plane bending fatigue test of the actual pipe

On the basis of the values estimated above, the actual pipe was subjected to a pulsating bending fatigue test. Figure 4.215 shows the relation between the stresses applied to various parts of the pipe and the number of cycles in a pulsating plane bending fatigue test of the actual pipe. When a pressing load of 2.0–4.0 tf is applied from the outside surface of the pipe, a tensile stress of 30–50 kgf/mm² maximum is induced at the inside surface of the pipe and the compressive stress is developed at the outside surface of the pipe in the vicinity of the pressed part. Under this stressing condition, therefore, cracks initiate at the inside surface of the pipe and propagate, but cracks do not initiate at the outside surface under the compressive stress. Accordingly, failure caused by this test does not agree with failure of the actual pipe.

One of the reasons for crack initiation and propagation from the outside surface of the pipe under the compressive stress and propagation is the effect of residual stress in the pipe. Since a residual tensile stress exists in the outside surface of the pipe, the outside surface of the pipe under the pulsating compressive stress is turned into a partially reversed stressing condition, including a tensile stress.

Figure 4.216 shows the results of measurement of residual stress in the circumferential direction of the pipe of the same steel grade as the pipe from which water leaked. A residual tensile stress of 10–20 kgf/mm² exists on the outside surface of the pipe. Figure 4.217 shows the stress induced in the outside surface of the pipe in cases where the pressing load is repeatedly applied from the outside surface of the pipe under the condition described above. It will be understood that the outside surface of the pipe is turned into a partially reversed stressing condition, including a

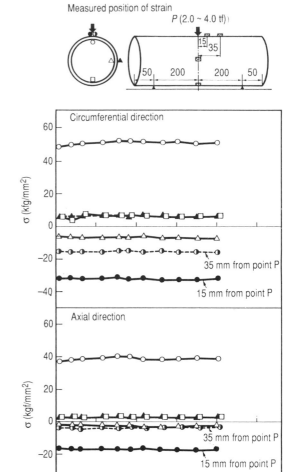

Figure 4.215 Measured values of stress at the surface of pipe

tensile stress, due to the existence of a residual tensile stress.

Figure 4.218 shows the cracked surface produced after a pulsating plane bending fatigue test of the actual pipe. Cracks initiated and propagated from the outside surface under the pressing load and from the inside surface. The condition is similar to that in the pipe from which water leaked. From the failure condition observed in the pipe from which water leaked and the results of the laboratory test, it is apparent that the cause of failure of the pipe is fatigue fracture caused during transportation.

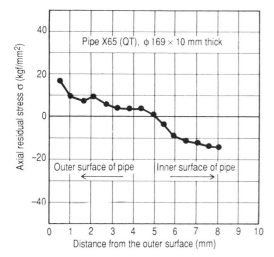

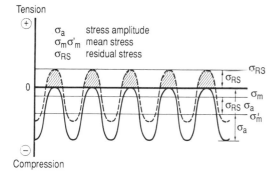

Figure 4.216 Axial residual stress distribution in pipe (turned out from outer surface)

Figure 4.217 Effect of residual stress as mean stress

4.11.5 Conclusion

According to the analysis of failure of the pipeline shipped to Alaska and the results of the laboratory test, it was found that a seamless pipe shipped from Japan in sound condition may be broken due to repetitive stress, depending on the conditions under which the pipe is stored on board ship. Although we have not touched on it here, corrosion by the action of sea-water may be one of the causes of deterioration of fatigue strength. Countermeasures should take this type of corrosion into account.

References

1. Yagi, A., Nishida, S., Higashiyama, H. *et al.* (1985) *Journal of the Iron and Steel Institute of Japan*, **71**, 1663, Tokyo
2. The Institute of Iron and Steel (ed.) (1973) *Iron and Steel Handbook*, Maruzen, Tokyo, p. 1503
3. Raju, J. S. and Newman J. C. Jr. (1979) *Engineering of Fracture Mechanics*, **11**, 817
4. Ishida, M. and Noguchi, H. (1982) *Transactions of the Japan Society of Mechanical Engineers*, **A48**, 607, Tokyo
5. Smith, F. W., Emery, A. F. and Kobayashi, A. S. (1967) *Journal of the Applied Mechanics*, p. 953
6. Nisitani, H. and Nishida, S. (1968) Preprints of 46th Meeting of the Japan Society of Mechanical Engineers, No. 198, 83, Tokyo

4.12 Failures starting from welds in machines and equipment [1]

As mentioned in Section 1.3.2, failures are often due to a wrong estimation of loading conditions, the use of improper assumptions in design calculations and inferior workmanship in assembling and installation. In other words, many failures are due to human neglect that can be avoided if careful investigation, analysis or inspection is carried out before the machines and equipment are made and used.

The failures discussed here are confined to those related to welded joints. Of the four cases discussed, three are fatigue failures. Conceivably, many failures of machine parts and structural members are related to welded joints (see Figure 1.4) [2]. In a welded joint separate pieces of metal are joined together. As such, welded joints are likely to produce defects, become the source of stress concentration and undergo various property changes. They are therefore generally more susceptible to failures than is the base metal. In addition, failed welded parts and members are difficult

Outer surface of pipe

Inner surface of pipe

Figure 4.218 Fracture surface of actual pipe in pulsating bending fatigue test

to substitute quickly. All things considered, therefore, the prevention of failures in welded joints is really important for assuring the smooth operation of machines and equipment.

Some examples of failures related to welded joints in machines and equipment are analysed in the following.

4.12.1 Case 1: crack in a rolling mill stand

(a) Condition of the crack
A crack occurred in a rolling mill stand that had been in use for about nine years. Figure 4.219 shows the rolling mill stand and the location of the crack being discussed. The crack, 210 mm long and 65 mm deep, occurred near the bottom end on the exit side of the stand. Macroscopically, the crack was substantially perpendicular to the surface of the stand. The stand was made of carbon steel casting according to the Japanese Industrial Standard JIS G 5101, SC42 whose chemical composition and mechanical properties are shown in Tables 4.44 and 4.45.

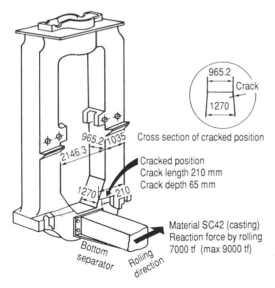

Figure 4.219   Schematic illustration of the rolling mill stand and position of crack

(b) Cause of the crack
Figure 4.220 shows the specimen taken for the purposes of investigation. Immediately after the crack was found, the surface of the stand was chipped away, to a depth of 10-plus mm, with a chisel as it did not look as deep as it really was. When the crack was not removed by chipping, it was decided to make a closer investigation. Figure 4.221 shows the macro appearance of the cracked (fracture) surface and the vertical section containing the crack. The cross section shown in Figure 4.221(b) was polished and etched with nital, as

Table 4.44   *Chemical composition* (wt %)

C	Si	Mn	P	S	Ni	Cr	Mo	Al
0.42	0.25	0.74	0.020	0.009	0.01	1.08	0.16	0.033

Table 4.45   *Mechanical properties**

$\sigma_{0.2}$ (kgf/mm^2)	σ_B (kgf/mm^2)	El (%)	ϕ (%)	$H_v(5)$
57.2	88.1	18.3	35.8	262

$\sigma_{0.2}$, proof stress; σ_B, tensile strength; El, elongation; ϕ, reduction in area.
* JIS No. 4, tensile specimen, mean value of $n=2$.

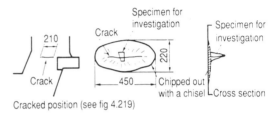

Figure 4.220   Method of obtaining a cut-out test specimen for investigation

shown in Figure 4.222. Although it is impossible to pinpoint as a result of chipping, the crack is estimated to have started in the nearby build-up welding whose trace can be found on the specimen.

The crack propagated initially somewhat diagonally and then, from midway, substantially perpendicularly to the surface. Another crack, 15 mm in length, was found near the estimated crack initiation point.

Figures 4.223 and 4.224 show the microstructures of the specimen at points A, B and C shown in Figure 4.222. The following conclusions may be derived from the observation of the photomicrographs.

1. The estimated crack initiation point shows a structure of deposited steel, suggesting the application of build-up welding
2. The base metal consists of ferrite and pearlite, with no signs of a significant defect
3. The crack seems to propagate mostly transgranularly

(c) Electron microscanning observation of the fractured surface
As is obvious from Figure 4.223 etc., the fracture surface of the specimen had rusted considerably. Therefore, the derusted surface was observed under a scanning electron microscope (SEM). Figure 4.225 shows an example of the striations observed at point S

(a)   (b)

Figure 4.221 Outer appearance of test specimen for investigation: (a) fracture surface, damaged by chipping with a chisel; (b) cross section including the cracked portion: top end, the surface chipped with a chisel; left end, the fracture surface, observed from the direction shown by the arrow above

in Figure 4.221(a). It is difficult to determine the type of the fracture surface because the area in the vicinity of the crack initiation point was chipped away. But the presence of striations characteristic of fatigue suggests that the crack was formed as a result of fatigue.

(d) Study by the fracture mechanics approach

Although not determined, it is said that the threshold of the stress-intensity factor K_{Ith} for JIS G 5101, SC42 carbon steel casting falls substantially within the range 20–28 kgf/mm$^{3/2}$ irrespective of the strength of the material [3]. If $K_{Ith} = 20$ kgf/mm$^{3/2}$, the crack on the rolling mill stand being discussed may be approximated to a surface crack on a semi-infinitely large plate because the thickness and width are sufficiently larger than the size of the crack [4].

$$K = 1.2 \ \sigma \sqrt{(\pi a)} \qquad (4.50)$$

If the rolling reaction force is 7000 tf, the nominal stress in the post corner region may be 3.0 kgf/mm [5]. From equation (4.50), therefore, $a = 9.8$ mm is derived. This means that no crack propagation occurs unless there is an initial defect approximately 10 mm in depth.

The presence of a defect about 10 mm deep having been postulated, the degree of propagation during the service period of nine years (the number of applied stress cycles $N = 1.39 \times 10^7$ times) was estimated and

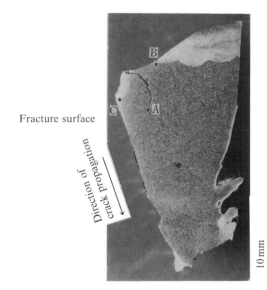

Fracture surface

Figure 4.222 Macroscopic structure in the vertical section (the left end is the fracture surface)

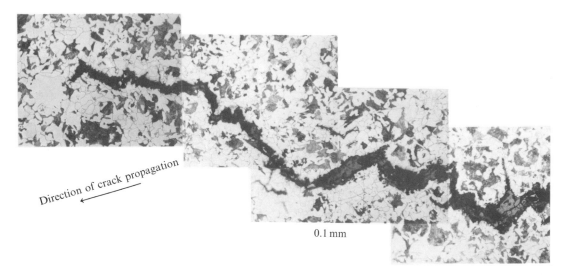

Figure 4.223   Microstructure of tip part of the crack (point A in Figure 4.222)

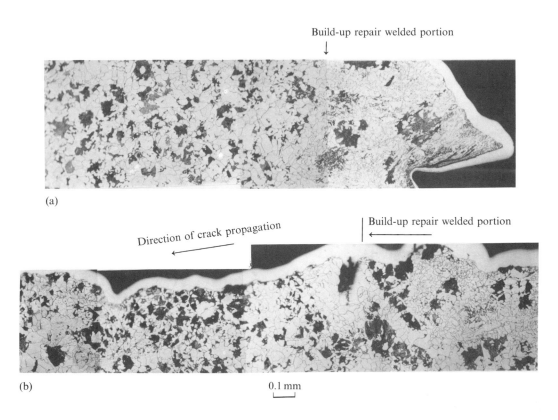

Figure 4.224   Microstructure of the built-up repair welded portion and the vicinity of crack initiation point: (a) point B of Figure 4.222; (b) point C of Figure 4.222

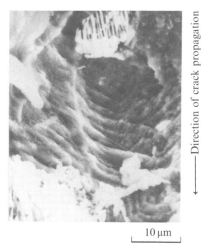

Figure 4.225 Striations observed at point S in Figure 4.221(a)

the value obtained was compared with the depth of the actual crack (approximately 65 mm).

Crack propagation in carbon steel for machine structural use according to JIS G 4051 S45C can be expressed as follows (see Section 3.3.2):

$$da/dN = 2.12 \times 10^{-12} \, \Delta K^{3.89} \qquad (4.51)$$

When equations (4.50) and (4.51) were integrated after substituting $N = 1.39 \times 10^7$, it became apparent that the crack grows to a length of only 12.1 mm. This limited growth was due to a level of stressing that was as low as 3.0 kgf/mm^2. When the same calculation was repeated by assuming an initial defect 20 mm deep, the depth of the propagated crack became 29.8 mm, which was quite close to the depth of the actual crack. In the quantitative estimation of fatigue crack propagation, it seems necessary to consider the influence of the applied stress [6,7] and the residual stress [8]. But this requirement is difficult to satisfy in practice.

(e) Summary and corrective measure

It is supposed that the rolling mill stand had an initial defect (approximately 10–20 mm deep; possibly an underbead cracking because carbon equivalent C_{eq} was as high as 0.806) formed by, for example, the build-up welding in manufacturing, which later grew into a fatigue crack under repeated application of stresses in service.

The easiest and surest remedy is to grind off the cracked region because the crack is, although quite deep, very small compared with the stand itself and, in addition, lies in a smooth area. An increase in nominal stress resulting from the grinding is substantially negligible.

As the nominal stress and the number of applied stress cycles to failure can be derived from the intervals between striations in the fracture surface, the reader is recommended to make further studies.

4.12.2 *Case 2: failure in an unloader equalizer*

(a) General description of the failure

Figure 4.226 is a schematic illustration of the unloader being discussed. The failure was found in a welded portion of the equalizer. Obviously, it was caused as a result of fatigue.

Figure 4.227 shows the profile of the equalizer and its fatigue-fractured portion. Having been in service for 1.5 years, the unloader was estimated to have been subjected to repeated stressing 24×10^4 times.

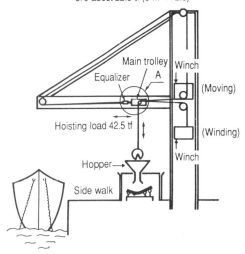

Figure 4.226 Outline of the unloader

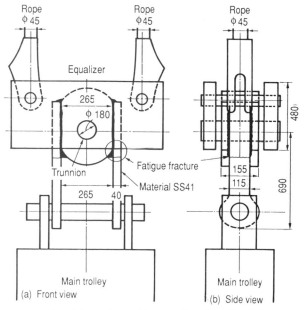

Figure 4.227 Dimensions of the equalizer (detail of A in Figure 4.226): (a) front view; (b) side view

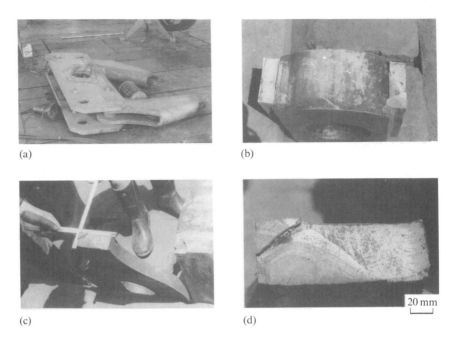

Figure 4.228   Fractured portion of the equalizer: (a) upper portions of equalizer; (b) outer view of the broken portions; (c) matching fracture surface of right-hand side of (b); (d) enlarged view of (c)

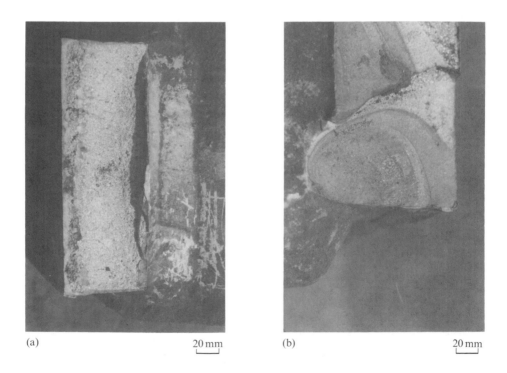

Figure 4.229   Fracture surface of the equalizer: (a) fracture surface of left-hand side of Figure 4.228(b); (b) fracture surface of right-hand of Figure 4.228(b)

(b) Cause of the fracture

Figure 4.228 shows the fractured portion of the equalizer. Starting in the weld on the right-hand side of the member, fatigue fracture propagated in the direction of the thickness and width as shown in Figure 4.228(b) (see also Figure 4.229(b)). Meanwhile, the left-hand side of the member shown in Figure 4.228(b) largely exhibits brittle fracture, with a very small amount of fatigue fracture that can be found with careful observation. Obviously, therefore, fracture occurred earlier on the right-hand side of the member shown in Figure 4.228(b).

Figure 4.230 shows a fracture surface observed under an SEM. The observation was made in a region with propagated fracture 15 mm away from the crack initiation point shown in Figure 4.228(b). Striations characteristic of fatigue are clearly perceived.

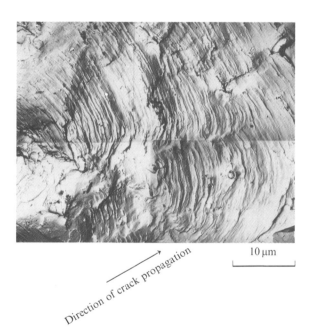

Figure 4.230 An example of the fracture surface observed by SEM (15 mm from the crack initiation point; see Figure 4.229(b))

The nominal static stress in the fractured portion is 4.6 kgf/mm^2. But the unloader's coefficient of dynamic load is quite large because the unloader swings when it hoists a bucket, moves and stops. Efficient operation of the unloader is required because shorter unloading time results in cost saving. If, therefore, the coefficient of dynamic load is lowered to 2, the nominal stress becomes 9.2 kgf/mm^2. Actually, the value of design fatigue limit σ_0 that the maker of the unloader chose seems to have been about 14.0 kgf/mm^2.

'The fatigue limit σ_s used in the strength design of a welded lap joint not finished by fillet welding is 4.7 kgf/mm^2 (with a fiducial probability of 95%) [9]. Although limited space does not permit a detailed discussion of the study results, it can be pointed out that the failure of the unloader equalizer is ascribable to the rather high stress and the failure to consider dynamic load in the fatigue strength design. The design fatigue limit is 14.5 kgf/mm^2 for 10×10^4 cyclic stressing and 7.9 kgf/mm^2 for 50×10^4. By interpolating, therefore, a design fatigue limit of 10.5 kgf/mm^2 can be derived for 24×10^4 cyclic loading.

(c) Summary and countermeasure

The failure in the unloader equalizer resulted from fatigue, probably because the fatigue limit chosen in the design was higher than the real one. The remedy for this problem is to keep the fatigue limit under 4.7 kgf/mm^2 by considering the influence of dynamic load.

4.12.3 Case 3: failure in a plating tank's welding joint

(a) Condition of the failure

Figure 4.231 shows the plating line being discussed. The failure occurred in a welded joint between the bottom plate and a side plate (both being of carbon steel for machine structural use according to JIS G 4051, S35C; see Figure 4.232) of its plating tank. The

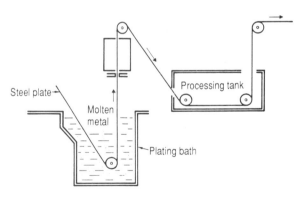

Figure 4.231 Schematic illustration of the plating line

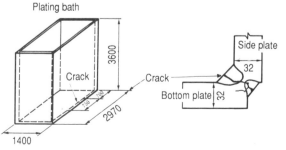

Figure 4.232 Position of crack in the plating bath

tank had been in service for a period of about three years, during which it had undergone temperature variations of approximately 150°C ten-plus times and those of 50–60°C tens of times. Normally, the tank was kept at 360 ± 5°C. The weld metal consisted of an alloy of lead and tin. The crack that occurred in the welded joint of the plating tank was approximately 750 mm long at its surface. When the failure was found, the crack was gouged and repaired by applying overlay welding because the production schedule did not permit a line shutdown. Although, therefore, the cause of the failure was not ascertained, the action taken may be inevitable with commercial lines. Engineers are often required to find appropriate remedies for problems where only circumstantial evidence is available.

(b) Cause of the failure

The careful non-destructive inspection given to the welded joints of the plating tank before it was put into use showed no sign of a crack at all. Therefore, weld cracks or other similar defects may safely be excluded from the causes of the failure. Judging from the service condition of the tank and the condition of the crack

found in it, fatigue due to thermal stress, embrittlement and cracking [10] under the influence of the liquid metal and lamination may be named as the cause of the failure. As thermal fatigue seemed to be the most likely cause, an attempt was made to study the fatigue characteristics of the fillet welded corner joint and to use the obtained data in the design of the plating tank. Improving the fatigue characteristics was considered to be effective for the prevention of other failures, too. Therefore, the fatigue characteristics of the fillet welded corner joint were evaluated using the same steel as that of the plating tank being discussed to determine the ideal profile of the joint as will be described hereunder.

(c) Fatigue characteristics of the fillet welded corner joint [11]

(i) Material, specimen and testing method
Table 4.46 shows the chemical composition and mechanical properties of the steel used in the test. Table 4.47 shows the types of specimen and welding conditions used in the test. Specimen A was analogous

Table 4.46 *Chemical composition and mechanical properties of test specimen*

| Kind of steel | Plate thickness | Chemical composition (wt %) | | | | | Mechanical properties | | | |
		C	Si	Mn	P	S	$\sigma_{0.2}$ (kgf/mm²)	σ_B (kgf/mm²)	El (%)	ϕ (%)
S35C	32 mm	0.38	0.27	0.78	0.020	0.013	35.5	62.6	32.8	55.0

$\sigma_{0.2}$, proof stress; σ_B, tensile strength; El, elongation; ϕ, reduction in area.

Table 4.47 *Type and welding conditions of test specimens*

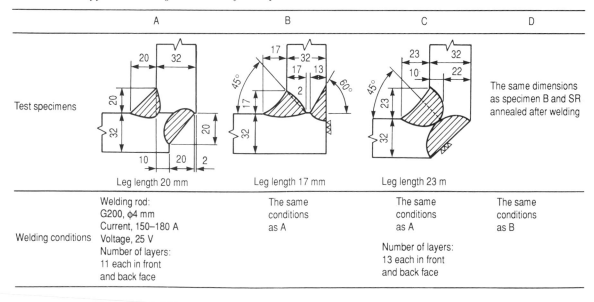

	A	B	C	D
Test specimens	Leg length 20 mm	Leg length 17 mm	Leg length 23 m	The same dimensions as specimen B and SR annealed after welding
Welding conditions	Welding rod: G200, φ4 mm Current, 150–180 A Voltage, 25 V Number of layers: 11 each in front and back face	The same conditions as A	The same conditions as A Number of layers: 13 each in front and back face	The same conditions as B

to the cracked welded joint being discussed, in which the amount of lapping at the fillet weld was made substantially equal to the plate thickness. Specimens B and C were prepared with a lap equal to the full thickness of the plate and with no lap at all, respectively. Specimen D was prepared by annealing specimen B at a low temperature to eliminate the influence of the residual stress created by welding. Figure 4.233 shows the profile of a test specimen.

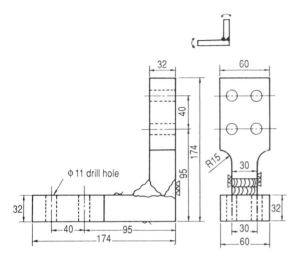

Figure 4.233 Dimensions of specimen B

A displacement controlled completely reversed test was performed on a torsional fatigue testing machine with a rated capacity of 1 tf m, with the specimen alternately bent back and forth while being rotated at a cyclic rate of 1650 cycles/min.

(ii) Test results and considerations
Figure 4.234 shows the S–N curves derived from the test results, in which stress amplitude σ_a is plotted as ordinate and the number of cycles to failure N_f as

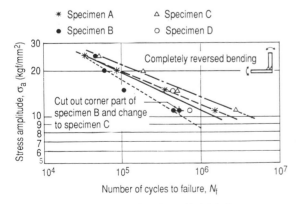

Figure 4.234 *S–N* curves for the welded joints

abscissa. In terms of fatigue characteristics, specimen C excelled best, followed by specimens A, D and B in that order. Although made to the same profile, specimen D exhibited a higher fatigue strength than specimen B. This seemed to be due to the removal of the residual tensile stress from the welded joint by low-temperature annealing.

It then became necessary to determine which of the types of joint and profile of the toe of the reinforcement of the weld was responsible for the difference in fatigue strength among the joints A, B and C.

Figure 4.235 plots the relationship between the radius R of the notch at the main initiating point of fracture and the flange angle θ. The relationship was determined by comparing the molybdenum compound of the weld prepared before the test and the main initiating point after fracturing. Shaped with a flank angle somewhat larger than that of joints A and B as can be seen in the figure, joint C was supposed to exhibit a lower fatigue strength. But the test result was reversed (see Figure 4.234). Therefore, the difference in fatigue strength among the different types of welded joint shown in Figure 4.234 proved unascribable to the profile of the toe of the reinforcement of the weld.

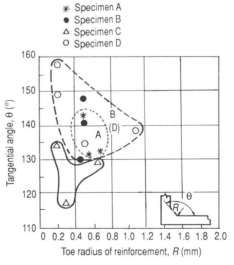

Figure 4.235 Comparison of shapes of toe of reinforcement of specimens

Next, stress analysis by the FEM (finite element method) was performed on different types of welded joint to clarify the relationship between their fatigue strength and stress concentration. The results of the analysis are shown in Figure 4.236.

As may be seen, stress was highest at the toe of the reinforcement of the weld on all types of welded joints, tapering off as the distance therefrom increased until the value became substantially constant at a distance of about 30 mm.

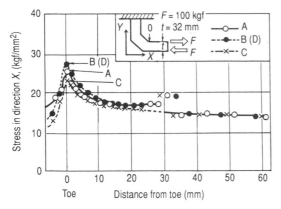

Figure 4.236   Comparison of stress distribution at toe of reinforcement for each specimen (by FEM analysis)

Table 4.48   *Relation between stress concentration factor by FEM at toe of reinforcement and fatigue strength*

Specimen	*Leg length (mm)*	*Fatigue strength at 10^6 cycles (kgf/mm²)*	*Stress con- tration factor by FEM, α*	*Ratio of α based on specimen C*
A	20	12.0	1.87	111
B	17	8.5	2.01	120
C	23	13.0	1.68	100
D*	17	11.0	2.01	120

* Annealed specimen B for stress relieving.

Figure 4.237 shows the relationship between the stress concentration factor α (the ratio of maximum stress to nominal stress) and leg length, while Table 4.48 shows the stress concentration factors determined by the FEM and fatigue strengths of the different types of welded joint. Joint B exhibited the highest stress concentration factor α at 2.01, followed by joint A at 1.87 and joint C at 1.68 in that order. This order agreed well with that of fatigue strength. As such, the difference in fatigue strength among the different types of joint may seem to be due to the difference in stress concentration in the different types of joint. But the influence of the leg length, which differs with the types of joint, is also included.

Table 4.49 shows the strains produced by bending stress near the toe of the reinforcement of the weld (5 mm away from the toe towards the base metal) in

Table 4.49   Comparison of strain due to nominal bending stress near each toe (5 mm distance) between specimen B and the specimen cut out at the outer corner

Specimen	Measured position by strain gauge	Strain due to nominal bending stress near the to (5 mm distance) × 10^{-6}	
		Nominal bending stress 12 kgf/mm²	Nominal bending stress 15 kgf/mm²
B	5 Strain gauge	625	830
As specimen B but with corner part cut out	5 Strain gauge	620	825

* mean value of $n = 2$

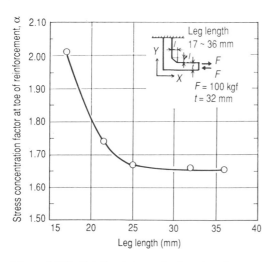

Figure 4.237   Relation between leg length and stress concentration factor α at toe of reinforcement (by FEM analysis)

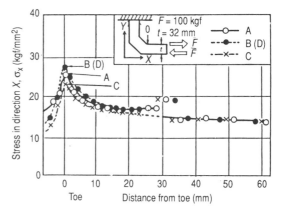

Figure 4.239   Comparison of stress distribution at toe of reinforcement for each specimen

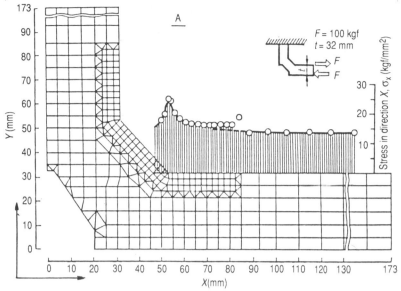

Figure 4.238 Stress distribution at toe of reinforcement for specimen A (by FEM analysis)

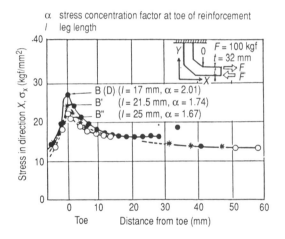

Figure 4.240 FEM analysis of change in stress distribution at toe of reinforcement produced by changing leg length

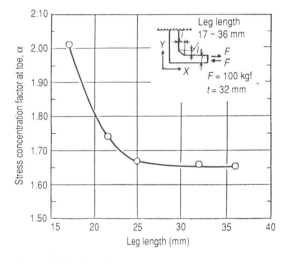

Figure 4.241 Relation between leg length and stress concentration factor (for specimen B)

joint B and another specimen prepared by removing the outer corner of joint B (which is apparently analogous to joint C). Under two different nominal bending stresses, the two specimens exhibited little difference in strain and fatigue strength near the toe of the reinforcement of the weld (see Figure 4.231). Thus, it may safely be concluded that the difference in the stress concentration factor α shown in Table 4.48 is not due to the type of welded joint.

The stress concentration factor at the toe of the reinforcement of the weld was analysed by the FEM while varying the leg length of joint B. Figure 4.237 shows an example of the analytical results obtained.

Figure 4.238 compares the stress distributions at the toe of the reinforcement of the weld on the individual joints, while Figure 4.239 shows how the stress distribution at the toe of the reinforcement of the weld varied with the changes in leg length. Figures 4.240 and 4.241 show the optimum leg length derived from the data obtained. As may be seen from the figures, the stress concentration factor became smaller with an increase in leg length, stopping to reduce further when the leg length became longer than 25 mm (approximately 80% of the plate thickness). All things considered, it may safely be concluded that the leg length mainly governs the fatigue strength of the fillet welded corner joint. The optimum leg length proved to be equal to approximately 80% of the plate thickness.

4.12.4 Case 4: failure in a hydraulic jack's reaction pedestal [1,2]

(a) Condition of the failure

When load was being applied with a hydraulic jack (with a rated capacity of 300 tf maximum) to remove a gear shaft, its lower reaction pedestal broke at 150 tf.

Figure 4.242 shows the appearance and breaking condition of the hydraulic jack. Reference number 1 designates the hydraulic jack, 2 and 3 reaction pedestals, of which the lower one 3 broke, and 4 supports (spaced 2030 mm apart).

Figure 4.243 shows the appearance of the broken reaction pedestal. Figure 4.244 shows the fracture surface of the broken reaction pedestal shown in Figure 4.243. An arc strike can be seen in the upper left-hand corner of the photograph shown at the right of Figure 4.240(b). Obviously, the fracture initiated from this arc strike.

Figure 4.242   Outer apupporting bar of reaction force (broken); 4, main frame (distance, 2030 mm)

Figure 4.243   Outer appearance of broken supporting bar

Figure 4.244 Fracture surface of supporting bar: (a) whole fracture surface; (b) crack initiation point of (a)

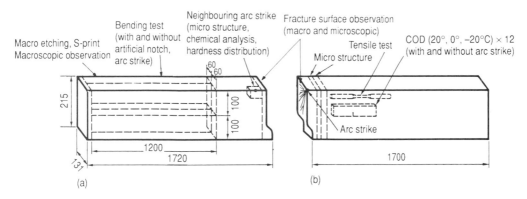

Figure 4.245 Investigation items and location of cut-out specimen

(b) Cause of the failure

Figure 4.245 shows the types of investigation performed and the locations where specimens were taken. The investigation comprised the determination of basic properties, observation of the fracture surface and study of breaking conditions.

(i) Basic properties of the material
The broken reaction pedestal was made of chromium molybdenum steel according to JIS G 4105, SCM440

made up of ferrite and pearlite. Evidently, the material was used as-rolled. Tables 4.50 and 4.51 show the chemical composition and mechanical properties of the material.

As shown in Figure 4.245, there was an adjoining arc strike in the vicinity (at a distance of about 60 mm) of the initiating point of the fracture. Figure 4.246 shows the adjoining arc strike. As may be seen in Figure 4.246(b), the arc strike exhibits an underbead crack which consists of a crack running along the heat-

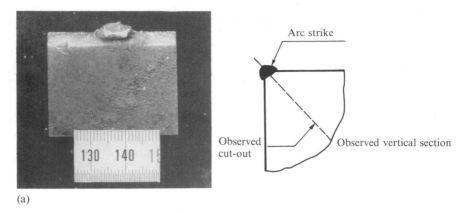

(a)

(b)   2 mm

Figure 4.246 Neighbouring arc strike (see Figure 4.244): (a) outer appearance; (b) vertical section of (a)

Table 4.50 *Chemical composition* (wt %)

C	Si	Mn	P	S	Ni	Cr	Mo	Al
0.42	0.25	0.74	0.020	0.009	0.01	1.08	0.16	0.033

Table 4.51 *Mechanical properties**

$\sigma_{0.2}$ (kgf/mm^2)	σ_B (kgf/mm^2)	El (%)	ϕ (%)
57.2	88.1	18.3	35.8

* JIS 4, ϕ 10 mm, mean value of two specimens.
$\sigma_{0.2}$, 0.2% proof stress; σ_B, tensile strength; El, elongation; ϕ, reduction in area.

affected zone of the weld metal and one intersecting substantially at right angles thereto, both being about 2.5 mm deep. It seems likely that a similar underbead crack occurred in the arc strike at the starting point of the fracture (see Figures 4.247 and 4.248).

Figure 4.249 shows the hardness distribution across the vertical cross-section of the adjoining arc strike (see Figure 4.246(b)). The distance from the surface of the deposited metal is plotted as abscissa. The heat-affected zone was hardened to about $H_v = 700$.

(ii) Observation of the fracture surface
Figure 4.247 shows an example of an electron micro-scanning observation of the starting point of the

500 μm

Figure 4.247   Results of observation by SEM of crack initiation point

Crack under bead   Crack under bead

Brittle fracture surface

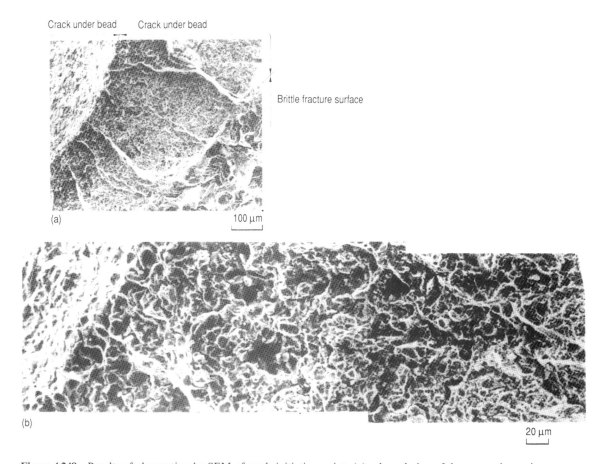

(a)   100 μm

(b)

20 μm

Figure 4.248   Results of observation by SEM of crack initiation point: (a) enlarged view of the arrowed area in Figure 4.247; (b) enlarged view of (a)

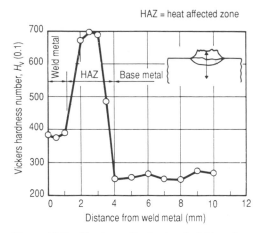

Figure 4.249 Hardness distribution (neighbouring arc strike). HAZ, heat affected zone

fracture. As may be seen, the fracture started in the heat-affected zone of the arc strike.

Figure 4.248 shows an enlarged view of a portion (pointed to by an arrow) of the area shown in Figure 4.247. As is obvious from Figure 4.248(b), many intercrystalline cracks were observed in the fracture surface, which seemed to correspond to the underbead cracking shown in Figure 4.244. The underbead cracks shown in Figures 4.247 and 4.246(b) seemed to be of substantially equal size.

(iii) Study of breaking conditions
First, specimens of the profile shown in Figure 4.250 were cut out of the fractured material. One of the

specimens was defect free, a second one was with an artificial notch (to a depth of 3 mm) resembling an underbead crack made at a corner in the middle of its length, and a third one was with a spot welding resembling an arc strike. The specimens were subjected to a three-point bending test. Figure 4.250 shows the relationship between the applied load and the resultant displacement determined by the test. The spot-welded specimen broke under a minimum load, exhibiting the smallest displacement. With the artificially notched specimen, the breaking load and displacement were 1.7 times and 4 times as large as those with the spot-welded specimen. The defect-free specimen was plastically deformed (by 95 mm) but did not fracture.

The influence of spot welding on fracture toughness was then studied using specimens for the crack opening displacement (COD) test according to ASTM E399. Figure 4.251 shows the profile of the COD test specimen (see Figure 4.245). Spot welding equivalent to an arc strike was applied at point A in Figure 4.251

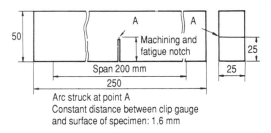

Figure 4.251 Dimensions of specimen

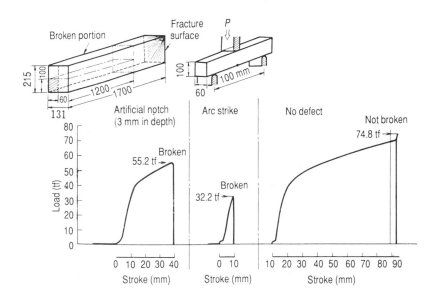

Figure 4.250 Simulation test results by three-point bending

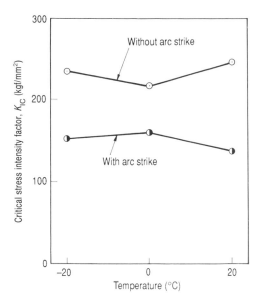

Figure 4.252 Test results by three-point bending

(with a G200 welding electrode 3.2 mm in diameter, a current of 150 A and a voltage of 220 V).

Specimens with and without spot welding were subjected to three-point bending tests at 20, 0 and −20°C to determine the critical COD value δ_c and the critical stress intensity factor K_{Ic}. The fracture surface on the broken specimens was observed under a scanning electron microscope, with attention focused on the region near the initiating point of the fracture. The results of the observation were studied together with the determined hardness distribution and the observation made under an optical microscope.

Figures 4.252–4.254 show examples of the observations obtained. Figure 4.252 shows the critical stress intensity factor K_{Ic} determined by the three-point bend tests. Figure 4.253 shows the fracture surface on the tested specimens. Figure 4.254 shows a photomicrograph of the fracture surface shown in Figure 4.253(b) observed under a scanning electron microscope. Obviously, the presence of spot welding served to considerably lower fracture toughness. As revealed by the observation of the fracture surface, ultimate fracture initiated at the tip of a fatigue crack substantially in the

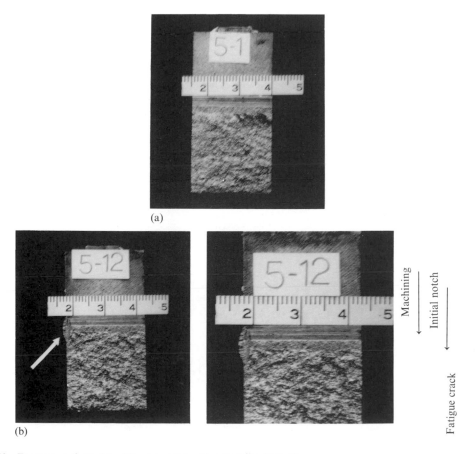

Figure 4.253 Fracture surfaces (a) without and (b) with arc strike (arrow)

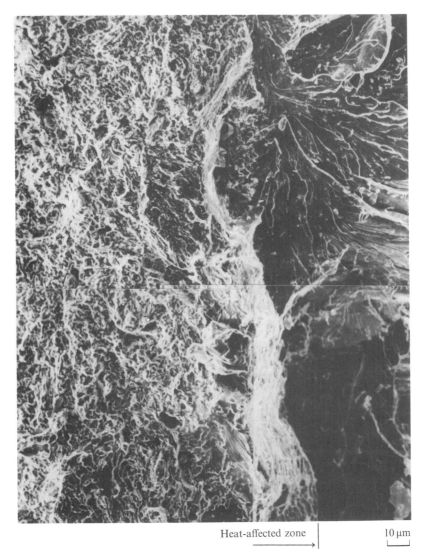

Heat-affected zone →   10 μm

Figure 4.254   Results of observation by SEM of crack initiation point (see Figure 4.253(b))

middle of the width of the specimen without spot welding. On the specimen with spot welding, ultimate fracture mainly initiated from near the boundary between the spot-welded region and the tip of a fatigue crack. With some types of base metal and microstructure, spot welding produces a considerable effect on fracture toughness. The structure of the specimen near the fracture initiating point seemed to have a considerable effect.

Next, the influence of arc strike on the fracture toughness of ten different types of rolled steel for general structures etc. (see Tables 4.52–4.54), using specimens of the profile shown in Figure 4.251. The test results obtained are shown in Table 4.55 and Figures 4.255 and 4.256.

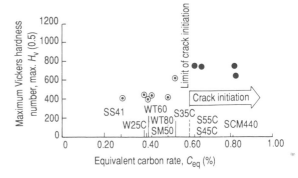

Figure 4.255   Relation of crack initiation limit in welds and maximum H_v

Table 4.52   *Materials used*

Kind of steel	Remarks	Thickness of plate (mm)	Heat treatment	Direction of cut-out
SS41	1–1 ~ 1–20	32	–	Rolling direction
S25C	2–1 ~ 2–20	32	–	ibid
S35C	3–1 ~ 3–20	32	–	ibid
S45C	4–1 ~ 4–20	32	–	ibid
S55C	5–1 ~ 5–20	32	–	ibid
SCM440	SCM–1 ~ SCM–12	131	as rolled	ibid
SCM440	M 4–1 ~ M 4–12	30 (where heat treated)	850°C × 60'OQ 600°C × 60'WQ	ibid
SM50	M 5–1 ~ M 5–20	35	–	ibid
WT60	W 6–1 ~ W 6–20	40	QT	ibid
WT80	W 8–1 ~ W 8–20	36	QT	ibid

QT, quenched and tempered.

Table 4.53   *Chemical composition (wt %)*

Kind of steel	C	Si	Mn	P	S	Ni	Cr	Mo	V	Al	B	Others
SS41	0.17	0.19	0.63	0.025	0.022	–	–	–	–	0.030	–	–
S25C	0.30	0.23	0.47	0.018	0.018	–	–	–	–	0.020	–	–
S25C	0.42	0.21	0.69	0.022	0.004	–	–	–	–	0.023	–	–
S45C	0.49	0.23	0.77	0.022	0.007	–	–	–	–	0.042	–	–
S55C	0.52	0.18	0.70	0.014	0.014	–	–	–	–	0.029	–	–
SCM440	0.42	0.25	0.74	0.020	0.009	0.01	1.08	0.16	–	0.033	–	–
SM50	0.18	0.43	1.31	0.025	0.009	–	–	–	–	0.031	–	–
WT60	0.11	0.23	1.25	0.014	0.003	0.16	–	0.13	0.036	0.059	0.0009	N 0.0033
WT80	0.11	0.23	0.87	0.013	0.005	0.79	0.50	0.44	0.037	0.058	0.0006	Cu 0.22

Table 4.54   *Mechanical properties**

Kind of steel	$\sigma_{0.2}$ (kgf/mm^2)	σ_{B} (kgf/mm^2)	El (%)	ϕ (%)
SS41	28.5	45.2	37.7	65.1
S25C	26.3	49.4	35.9	57.8
S35C	34.5	62.1	35.0	53.0
S45C	34.3	71.1	22.2	40.9
S55C	36.7	71.3	20.2	41.2
SCM440	57.2	88.1	18.3	35.8
SCM440	72.3	95.1	22.7	65.7
SM50	35.7	56.0	39.0	72.6
WT60	58.5	69.2	28.6	76.9
WT80	71.2	80.6	27.0	74.5

* JIS 4, ϕ10 mm, mean value of two specimens.
$\sigma_{0.2}$, 0.2% proof stress; σ_{B}, tensile strength;
El, elongation; ϕ, reduction in area.

Table 4.55   *Crack initiation limit in welds*

Kinds of steel	Heat treatment	C_{eq} (WES)	Crack initiation
SS41	–	0.283	○Not cracked
S25C	–	0.388	○Not cracked
S35C	–	0.544	○Not cracked
S45C	–	0.628	●Cracked
S55C	–	0.645	●Cracked
SCM440	as rolled	0.809	●Cracked
SCM440	850°C × 60 OQ 600°C × 60'WQ	0.809	●Cracked
SM50	–	0.416	○Not cracked
WT60	QT	0.403	○Not cracked
WT80	QT	0.498	○Not cracked

$$C_{\mathrm{eq}} = C + \frac{Si}{24} + \frac{Mn}{6} + \frac{Ni}{40} + \frac{Cr}{5} + \frac{Mo}{4} + \frac{V}{14}$$
(WES standards) QT, quenched and tempered.

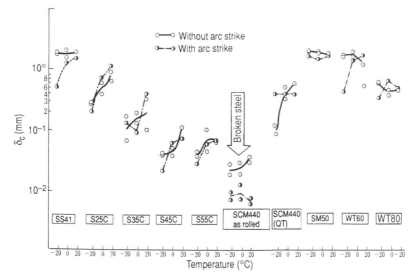

Figure 4.256   Relation of kind of steel and critical COD value δ_c (δ_c is calculated according to BS 5762)

4.12.5 Summary

The findings obtained from the case studies on the four failures discussed above are summarized in the following.

(a) Case 1

The crack initiated from a weld crack formed in the build-up metal welded when the rolling mill stand in question was made. A fatigue crack then occurred from this initiation point and propagated and grew into a macro crack.

(b) Case 2

The fatigue failure in the unloader equalizer initiated from its fillet welded lap joint. The fatigue failure proved ascribable to the dynamic load acting on the equalizer and the over-estimation of the design fatigue limit for the unfinished fillet welded lap joint.

(c) Case 3

The failure of the plating tank occurred in the welded joint between the bottom and side plates. There was not enough time to make a detailed investigation to determine the real cause of the failure. From the service condition of the plating tank and the crack found in it, fatigue due to thermal stresses seemed to be most responsible. Accordingly, the fatigue characteristics of the fillet welded corner joint were studied. The study showed that leg length chiefly governs the fatigue strength of the fillet welded corner joint and that the optimum leg length was about 80% of the thickness of the welded plate.

(d) Case 4

The failure in the reaction pedestal of the hydraulic jack was of a brittle fracture type and started from an arc strike formed during tack welding or other operations. Arc strike can produce a considerable effect on fracture toughness depending on the type and structure of the base metal. With as-rolled chromium molybdenum steel according to JIS G 4105, SCM440 in particular, the structure of a very thin layer in the heat-affected zone of the arc strike or of a region near the initiation point of the fracture has a considerable influence.

(e) Follow-up test on Case 4

This showed that an arc strike formed on various types of structural steel causes underbead cracking and lowers fracture toughness when carbon equivalent C_{eq} is equal to or greater than 0.60%.

References

1. Nishida, S. (1983) *Journal of the Japan Welding Society*, **2**, 22, Tokyo.
2. Nishida, S., Urashima, C. and Masumoto, H. (1981) In *Proceedings of the 26th National Symposium on Strength, Fracture and Fatigue*, p. 91, Sendai, Tokyo
3. Kitagawa, H., Nisitani, H. and Matsumoto, T. (1976) *Transactions of the Japan Society of Mechanical Engineers*, **42**, 996, Tokyo
4. Burdekin, F. M. and Dawews, M. G. (1971) *Journal of the Japan Welding Society*, **40**, 97, Tokyo

5. Hashiguchi, T., Marumoto, S., Yotsuya, K. *et al.* (1981) *Seitestu Kenkyu*, No. 305, 70, Tokyo

6. Kikukawa, M., Jono, M. and Kondo, Y. (1983) *Transactions of the Japan Society of Mechanical Engineers*, **49**, 278, Tokyo

7. Yamada, K. (1985) *Proceedings of the Japan Society of Civil Engineering and Construction*, **2**, 25, Tokyo

8. Nishida, S., Urashima, C. and Masumoto, H. (1983) *Journal of the Society of Materials Science*, **2**, 57, Kyoto, Japan

9. Ohta, S. and Sejima, I. (1978) *Design and Criteria of Welded Structures*, Sangyo Tosho, Tokyo, p. 125

10. Kikuchi, M. (1983) *Journal of the Japan Society of Mechanical Engineers*, **86**, 237, Tokyo

11. Urashima, C., Nishida, S. and Masumoto, H. (1982) *Journal of the Japan Welding Society*, **51**, 599, Tokyo

12. Nishida, S., Urashima, C. and Masumoto, H. (1980) Preprints of the Japan Society of Mechanical Engineers, No. 800–8, 85, Tokyo

4.13 Failures in rolls

Steel products other than cast and forged steels are made by rolling. Rolling operations can be roughly divided into hot rolling and cold rolling. Rolling mills are the main equipment used in both practices and rolls play a leading role in them.

This section deals with the findings obtained from macro analyses of failures that occurred in work rolls of downstream stands in the finishing train of hot strip mills that can be regarded as a typical example of hot rolling mills [1].

With the improvement in the performance of hot strip mills and the quality of rolled products, the quality requirements for the rolls used with them have become severer than before [2]. For example, the hot finisher's work rolls (hereinafter called the HFW rolls) discussed herein are made of high-alloy mottled cast iron because the iron has a high wear resistance, high breaking strength and stable surface. In order to satisfy the sophisticated functional requirements of the most modern mills, however, further improvement in wear resistance, in particular, is required.

In general, however, wear resistance conflicts with breaking strength in many respects. When increasing wear resistance, therefore, it is necessary to take care to minimize the impairment of breaking strength. To help such improvement, examples of failures in HFW rolls will be briefed first. Clinical analysis will then be made of cracking and spalling characteristic of HFW rolls, with consideration given to the mechanism of their occurrence and the preventive measures to be taken against them.

4.13.1 Types and shapes of failures in HFW rolls

About 50% of the HFW roll is consumed by wear in rolling, 15–20% by grinding to remove cracks and other failures and the rest by ordinary dressing. Figure 4.257 shows different types of failure occurring in the barrel of HFW rolls. Table 4.56 shows the relationship among the types of roll failures, roll materials and rolling conditions. Failures initiating from within rolls are mostly ascribable to defective casting, insufficient strength, excess residual stresses or the like except when they are not sufficiently cooled or otherwise properly operated (see Figure 4.257(b)–(c)). Under abnormal rolling conditions, such as a cobble, failures similar to those occurring during normal rolling are sometimes observed (see Figure 4.257(e) and (f)).

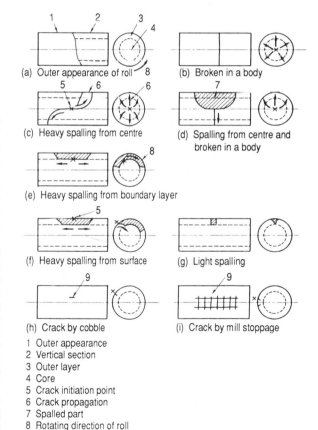

(a) Outer appearance of roll

(b) Broken in a body

(c) Heavy spalling from centre

(d) Spalling from centre and broken in a body

(e) Heavy spalling from boundary layer

(f) Heavy spalling from surface

(g) Light spalling

(h) Crack by cobble

(i) Crack by mill stoppage

1 Outer appearance
2 Vertical section
3 Outer layer
4 Core
5 Crack initiation point
6 Crack propagation
7 Spalled part
8 Rotating direction of roll
9 Surface crack

Figure 4.257 Schematic patterns of breakage in a roll body: (a) outer appearance of roll; (b) break in body; (c) heavy spalling from centre; (d) spalling from centre and break in body; (e) heavy spalling from boundary layer; (f) heavy spalling from surface; (g) light spalling; (h) crack by cobble; (i) crack by mill stoppage

Table 4.56   *Relation between breakage and cause of roll*

Rolling condition	Failure	Defect	Strength	Residual stress	Thermal stress	Rolling stress	Reference
				Cause of failure			
Normal	Broken in the body	○	○	○	○	–	Figure 4.257(b)
	Heavy spalling from centre	○	○	○	△	○	Figure 4.257(c),(d)
	Heavy spalling from boundary layer	○	○	○	△	○	Figure 4.257(e)
	Heavy spalling from surface	○	–	–	–	△	Figure 4.257(f)
	Light spalling from surface	○	–	–	–	△	Figure 4.257(g)
Abnormal	Heavy spalling*	–	–	–	○	○	Figure 4.257(f)
	Light spalling*	–	–	–	○	○	Figure 4.257(g)
	Crack by cobble	–	–	–	○	○	Figure 4.257(h)
	Crack by mill stoppage	–	–	–	○	–	Figure 4.257(i)

* Broken in the body occasionally; see Figure 4.258. ○, large effect; △, small effect; –, small or no effect.

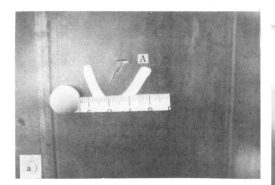

A ↑

Figure 4.258   Break in body due to spalling: (a) crack A, initiation point of fracture; (b) break in body due to spalling that initiated from crack A

Causes of individual failures, including whether the aforementioned analogous failures are ascribable to rolls themselves or rolling conditions, will be discussed in the following.

(a) Break-off of roll barrel

The break-off of roll barrels is also known as thermal breaking, manifesting a characteristic fracture pattern spreading from the centre to the periphery of rolls. The break-off of roll barrels sometimes results from the surface crack growing inwards or secondarily from a major spalling (see Figure 4.257(d)).

(b) Heavy spalling

When a heavy spalling resulting chiefly from the quality of the roll itself forms a fracture exceeding 100 mm in width, a beach mark spreading from the inside to the surface appears at the fracture surface (Figure 4.258). With a heavy spalling resulting from a casting defect at the surface, however, the fracture propagates from the surface to the inside (Figures 4.259 and 4.260). Even with heavy spallings that appear to have initiated from the inside, it may safely be considered that the actual initiation is at the surface when the fracture surface contains a path of propaga-

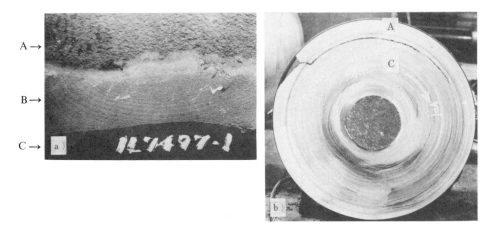

Figure 4.259   Heavy spalling from boundary layer: (a) fracture surface due to spalling: A, core; B, outer layer with beach mark; C, rolling surface; (b) transverse section of spalling

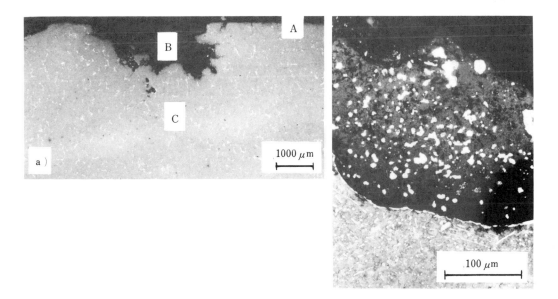

Figure 4.260   (a) Cast defect as initiation of spalling (etched with picric acid). (b) Magnified cast defect: A, rolling surface, B, defect (Al, Si, etc. detected); C, base metal of roll

tion from the surface (Figure 4.261). Figure 4.262 shows a case in which a larger crack opens up inside, as revealed when the surface layer is artificially removed (Figure 4.262(c)), under the surface that shows only a small crack (Figure 4.262(a)).

(c) Light spalling
A light spalling initiating at the surface of a roll, in a region free of casting defect, is considered to have grown from a cobble-induced crack discussed later under repeated application of the rolling load [3].

Because of its characteristic shape, the grip crack can be readily distinguished from other types of cracks (Figure 4.257(i)). The grip crack seldom brings about spalling. The crack that leads to spalling (Figure 4.263) is considered to result from a cobble, and a theoretical analysis to support this hypothesis has already been made [3].

4.13.2 Clinical analysis of spalling and cracks

The rolls analysed were the 80 rolls of high-alloy mottled cast iron that had been investigated during a

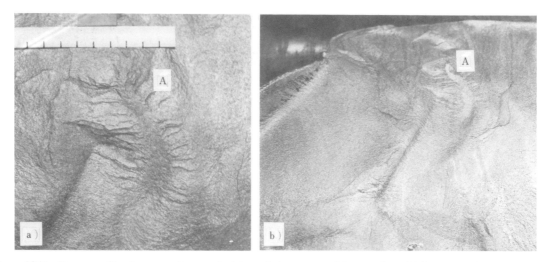

Figure 4.261   Heavy spalling from a surface crack: (a) crack A propagated from surface; (b) fracture surface of spalling initiated from crack A

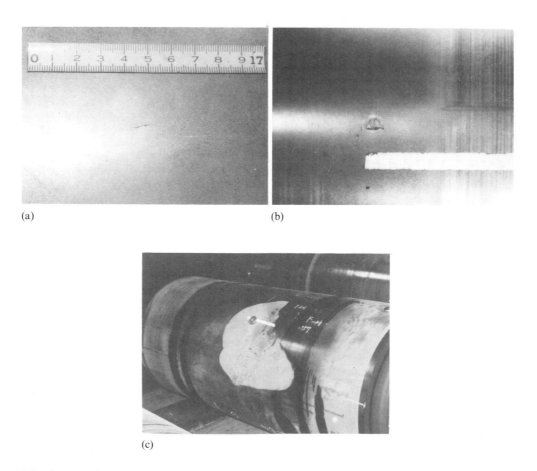

Figure 4.262   An example of light spalling apparently like a crack: (a) initiated crack; (b) crack artificially peeled off; (c) cut-off surface layer

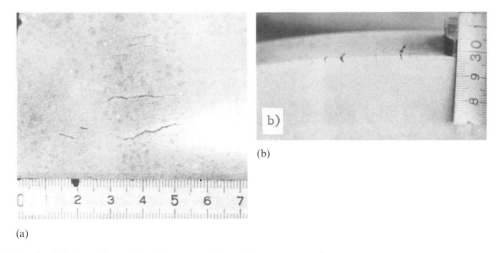

(a)

(b)

Figure 4.263   Crack induced by cobble: (a) outer surface; (b) transverse section

period of eight years. The rolls were confined to those where the makers had requested an investigation of the causes of failures and the resulting damage. There may be many failures that are ascribable to problems in the rolling operation, but these failures were not included in the scope of the analysis. Also, the three failures that were attributed to roll defects were not included. The above limitation should be considered in interpreting the analytical results to be described in the following. In making an analysis, an opening not larger than 1 mm was defined as a crack and a larger one as a light spalling. A spalling whose maximum projected axial length exceeds 100 mm was defined as a heavy spalling and one under 100 mm as a light spalling.

(a) Shape and characteristics of fractures

(i) Shape in external appearance
Figure 4.264 shows the classification of the external appearance of fractures by their shapes. Cracks were most extensively found, followed by light and heavy

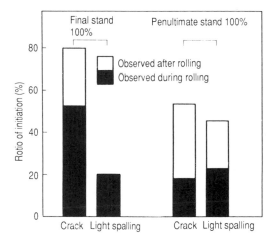

Figure 4.265   Outer appearance of crack and light spalling

spallings, in that order. Most of them were found during rolling. On the second last stand, however, those found during rolling were less than 50% (Figure 4.265).

(ii) Internal shape of fractures
Figure 4.266 shows the difference between the external and internal shapes of fractures. On the last stand, cracks and light spallings accounted for 80 and 20% of the fractures discovered, with only 50% of the cracks propagating inwards. On the second last stand, however, most of the fractures (over 90%) were light spallings. The reason for this will be discussed later in paragraph (d).

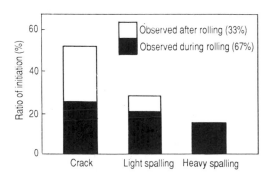

Figure 4.264   Outer appearance of failures (where a crack coexists with spalling, both are counted)

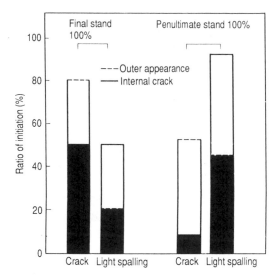

Figure 4.266   Difference in features between outer appearance and actual state of failure

(iii) Direction of propagation
Of the 25 instances of spalling, 23 propagated in the direction opposite to and one in the same direction as the rotating direction of the rolls.

(b) Location of fractures
Figure 4.267 shows the location of the cracks and spallings discovered. Quite naturally, many of them occurred in the middle portion.

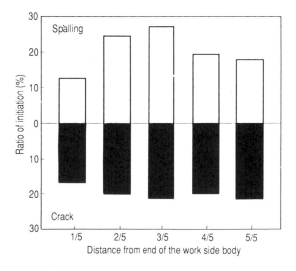

Figure 4.267   Fracture position of crack and spalling (WS, at the end of work side)

(c) Size of failures
Figure 4.268 shows the distribution of the width of spallings. The majority of the failures were light spallings under 100 mm in width extending in the direction of the roll axis, with a few exceeding 500 mm. More than 60% of the cracks were in the range 5–14 mm, with the shortest one at 3 mm (Figure 4.269).

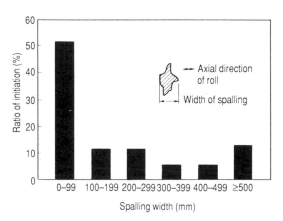

Figure 4.268   Distribution of spalling width

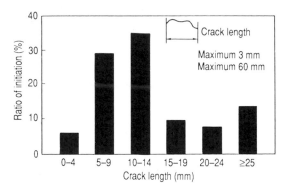

Figure 4.269   Distribution of crack length

(d) Depth of fractures
Figure 4.270 shows the distribution of the depth of fractures. Deeper fractures were found on the second last stand than on the last. This may be due to the difficulty in detecting the influence of roll defects on the surface of the strip, the higher likelihood of passing strips with unremoved cracks and the greater rolling load applied on the second last stand. While the depth of the cracks averaged 6.5–7 mm, light and heavy spallings were approximately three times and six times deeper, respectively (Figure 4.271). These data show the importance of the early discovery of defects.

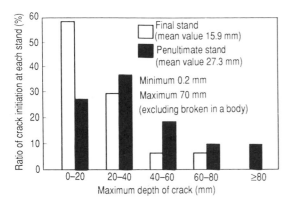

Figure 4.270   Distribution of crack depth

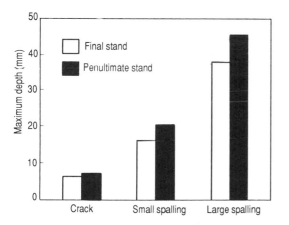

Figure 4.271   Maximum depth of crack and spalling

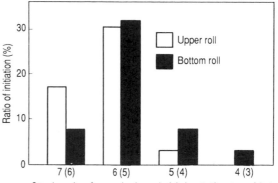

Figure 4.272   Relation between ratio of crack initiation and stand number

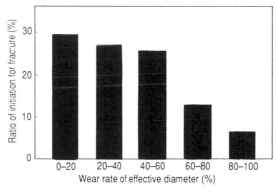

Figure 4.273   Relation between wear rate of effective diameter and ratio of initiation for fracture

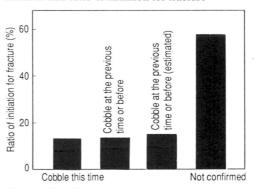

Figure 4.274   Relation between time of cobble and ratio of initiation for fracture (cobble at the previous time or before)

(e) Stand causing roll fractures

Figure 4.272 shows the relationship between the incidence of roll fractures and the stand number. The stand number is plotted as abscissa, with the figures in parentheses indicating the stand numbers on 6-stand mills. The incidence of roll fractures, including those conditioned in the rolling mill plants themselves, is highest on the last stand. The stands closer to the last stand proved more susceptible to cobble-induced troubles. The difference between the top and bottom rolls is not clear.

(f) Diameter of fractured rolls

Figure 4.273 shows the relationship between the ratio of consumption in effective roll diameter and the incidence of roll fracture. Obviously, most (about 80%) of the fractures occurred before the rolls had reduced to 60% of their effective diameter. These data seems to show the influence of the individual stands shown in Figure 4.272 as the HFW rolls are usually shifted from upstream to downstream.

(g) Relationship with cobbles

Figure 4.274 shows the relationship between the time at which cobbles occurred and the incidence of fractures. The fractures whose causes were ascertained as cobbles from operation records and other data were 27%. Their ratio did not exceed 50% even when estimated causes were included.

Table 4.57   *Main causes of initiation of cracks and spalling*

Features of failure		Causes of initiation of cracks and spalling			
Items	Results	Defect	Cobble	Contact fatigue	References
Outer appearance	Crack > spalling	○	○	○	Figure 4.264, Figure 4.265
Internal damage	Spalling > crack	○	○	○	Figure 4.266
Propagating direction of spalling	Opposite to rotational direction	○	○	○	
Position of spalling	Uniform distribution in axial direction	○	○	×	Figure 4.267
Position of crack	Uniform distribution in axial direction	○	○	×	Figure 4.267
Size of spalling and crack	Spalling: less than 100 mm Crack: less than 10–14 mm	–	–	–	Figure 4.268, Figure 4.269
Maximum depth of crack	Less than 20 mm	–	–	–	Figure 4.270
Stand number	Stand: 6 > 7 > 5 > 4	–	△	×	Figure 4.271
Upper roll and lower roll	No difference	○	○	○	Figure 4.272
Diameter of roll	Major diameter > minor diameter	–	○	×	Figure 4.273

○, large relation; △, medium relation; ×, small relation; –, no relation.

4.13.3   Consideration of analytical results

(a) Production mechanism of spalling
Table 4.57 summarizes the characteristics of spalling. The relationship between the analytical results and main causes, such as roll defects, cobbling and rolling contact fatigue, was studied. As roll defects can be found by colour checking and microscopic observation, spalling on sound HFW rolls should be ascribed to the initiation and propagation of cobble-induced cracks, as shown in Table 4.57.

(b) Prevention of spalling
Since it is difficult to eliminate cobbling from the rolling operation, the preventive measure against spalling must take it into consideration.

(i) Roll materials
Studies have been made on the use of roll materials that prevent the occurrence or propagation of cracks [2,4]. Also, attempts have been made on the elimination of roll defects and the improvement of their detection accuracy [2].

(ii) Maintenance of rolls
It is essential to find and remove cracks as thoroughly and early as possible. This goal has been almost achieved by applying a high-precision eddy-current flaw detection test every time rolls are dressed.

References

1. Sano, Y. and Kimura, K. (1985) *Journal of the Iron and Steel Institute of Japan*, **71**, 97, Tokyo
2. Sano, Y., Sugiura, Y., Eda, T. and Hirata, K. (1985) *Hitachi Hyoron*, **67**, 303, Tokyo
3. Sekimoto, Y. (1970) *Transactions of the Iron and Steel Institute of Japan*, **10**, 341, Tokyo
4. Sano, Y. (1982) Preprints of the 81st Symposium of the Japan Society for Technology of Plasticity, Kyushu Branch of the Japan Society for Technology of Plasticity, Kitakyushu, Japan

4.14   Failures in rail joints

As is well known, rails play a very important and indispensable role in railways. Rails too often fail. Typical rail failures are the wear and rolling contact fatigue of their head and the fatigue of their web and base, which are often accelerated in corroding environments. In ordinary railways, most rail failures occur in their joints, which can be classified as failures in bolt holes, upper and lower fillets and a combination of bolt holes and fillets. They are all fatigue failures starting from rail defects and notches [1].

One of the typical failures in rail joints is known as the upper fillet crack (technically the failure should be called the crack in the upper fillet of a rail, but the upper fillet crack is preferred here for its plainness). This type of failure has been found, mainly in underground railways, since about 1967. This section deals with the upper fillet crack as a typical example of failures in rail joints. Investigation of its cause, a laboratory reproduction test and preventive measures will be discussed [2,3].

4.14.1 Characteristics of the upper fillet crack

Figure 4.275 shows an example of an actual upper fillet crack of a rail. The upper fillet crack has the following characteristics:

1. Cracks occur at the upper fillet of rails that are in contact with joint plates.
2. Cracks grow towards the head substantially concentrically, at an inclination of about 60–70° with respect to the longitudinal direction of the rails.
3. Cracks frequently appear in heat-treated (quenched and tempered) rails (hereinafter called head-hardened rails).

4. Cracks frequently occur where joint plates are fastened with two bolts.
5. Cracks occur more frequently on the gauge corner side (with which the flange of the wheels comes into contact) of both inner and outer rails at curves.
6. Cracks occur frequently at unlubricated rail joints but rarely at lubricated ones.

4.14.2 Investigation and analysis of the upper fillet crack of rails

(a) Items investigated
The following investigations and analyses were made on the material of rails and their fracture surface:

1. Chemical analysis
2. Sulphur printing and macrostructure
3. Hardness distribution
4. Microscopic observation of cross section near the initiating point of cracks
5. Measurement of residual stress
6. Electron microscanning observation of the fracture surface.

(b) Investigation results and consideration
The following is the briefing of the investigation results obtained and the consideration based on them.

(i) Chemical analysis
Table 4.58 shows the results of the chemical analysis, whereas Figure 4.276 shows the distribution of hardness in the rail head. With nothing abnormal observed, the rail appeared to be made of a satisfactory material.

Figure 4.275 Example of crack at upper fillet (50N; head-hardened rail; total passing tonnage, 38.2×10^6 tf)

Table 4.58 Chemical composition of the rail

					(mass %)
	C	Si	Mn	P	S
JIS specification	0.60–0.75	0.10–0.30	0.70–1.10	< 0.035	≤ 0.040
Analysed value	0.68	0.13	0.86	0.019	0.020

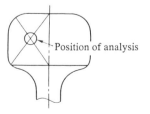

Position of analysis

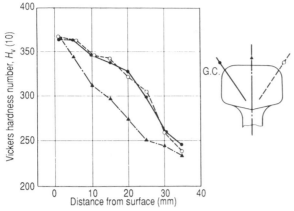

(a) Hardness distribution of rail head in transverse section

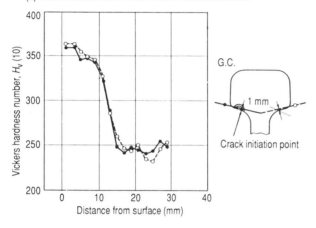

(b) Hardness distribution near the crack initiation point in transverse section

Figure 4.276 Hardness distribution of rail head: (a) in transverse section; (b) near the crack initiation point in transverse section

(ii) Microscopic observation in the vicinity of crack initiation

Figure 4.277 shows an example of a microscopic observation made across the vicinity of crack initiation. Although the photomicrograph shows only one fine crack, it was confirmed that many other cracks were present nearby. The fine cracks, mostly not longer than 100–200 mm, proved to be 'post-natal' ones that occurred in service.

The fine cracks were limited to where the rail was in contact with its joint plate. An accretion of iron oxide, known as 'cocoa', was found at the main starting point of the cracks. From these facts, the fine cracks may safely be considered to have resulted from fretting fatigue. For fretting fatigue, it is recommended to refer to the cited library [4].

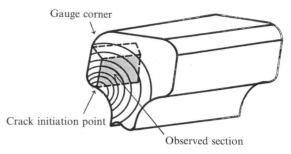

Figure 4.277 Small cracks at upper fillet (vertical section)

(iii) Measurement of residual stress

The fracture of the rail occurred at a point 270 mm away from its end. A strain gauge was affixed at a middle point 135 mm away from the fracture surface to measure the residual stress by the intercept method. The results obtained are shown in Figure 4.278. A longitudinal residual tensile stress of 15 kgf/mm² was found at the upper fillet where the fracture had initiated. Usually, there is a residual tensile stress of about 20 kgf/mm² at the upper fillet of a heat-treated rail (1981, unpublished data). Therefore it may safely be considered that there had originally been as much residual tensile stress in the fractured rail being discussed. The somewhat lower measurement may be due to the short length of the specimen, from which some of the original residual tensile stress must have escaped. Anyway, the residual tensile stress at the upper fillet of a rail is considered to have a function to accelerate the initiation and propagation of fatigue cracks. Strictly speaking, residual tensile stress accelerates the propagation rate of fatigue cracks. But the above expression in the text was chosen because the occurrence of a visually observable crack (the initiation of a crack) involves the process of its propagation too.

(iv) Observation of fracture surface

As is obvious from observation of the macrostructure shown in Figure 4.277, the fracture surface showed a

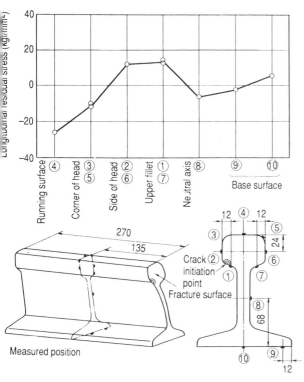

Figure 4.278 Residual stress in longitudinal direction of rail

beach mark (a shell-like pattern) characteristic of fatigue fracture, showing that the fracture was due to fatigue. Figure 4.279 shows a region near the initiation point of the crack observed under a scanning electron microscope. Although not clearly seen because of the secondary damage caused by the rubbing of fracture surfaces, the initiation point retains the feature of the fracture surface of fatigue failure. Inclusions and other metallurgical defects were not found at the initiation point.

All things considered, the upper fillet crack was not due to the quality and metallurgical defects of the material but due to fretting fatigue that often occurs where a joint plate contacts a rail. Because this type of crack occurs more frequently in the head-hardened rails, it seemed that there was a longitudinal residual tensile stress of 15–20 kgf/mm² at the upper fillet of the rail, which served as a factor to accelerate fretting fatigue.

4.14.3 Laboratory reproduction of an upper fillet crack and study of influencing factors

(a) Reproduction test of an upper fillet crack

Figure 4.280 shows how an upper fillet crack of a rail was reproduced. Two 50N head-hardened rails were joined together with an ordinary joint plate (with four bolt holes) bolted with a torque of 5000 kgf cm). The

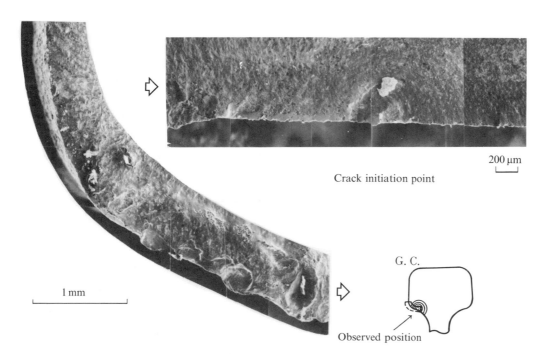

Crack initiation point

200 μm

1 mm

G. C.

Observed position

Figure 4.279 Fracture surface near the crack initiation point observed by SEM

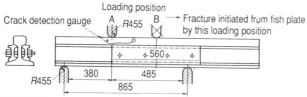

Loading position

Crack detection gauge   A   *R455*   B   → Fracture initiated from fish plate by this loading position

| 380 | 485 |

♦ 560 ♦

R455

865

Rail: 50 N heat treated (HH) rail

Normal fish plate connection with four holes (fastening torque 5000 kgf–cm)
Pulsating three-point bending by the fatigue tester with capacity of 150 tf
Base tension span 865 mm
Repeated load P_{max} = 20 tf P_{min} = 6.5 tf
Frequency 500 Hz
Crack detection gauge: aluminum foil

Figure 4.280   Test conditions

specimen thus prepared was subjected to a three-point bend test.

Figure 4.281 shows the appearance and fracture surface of the detected cracks. Macroscopically, cracks were found at three points: near the joint plate, at two points of the upper fillet and just above the bolt holes. Microscopically, however, several fine cracks, such as those shown in Figure 4.282 were found at the upper fillet in the vicinity of the joint plate. Powdery iron oxide known as cocoa was observed at their initiation points. Showing these features, the cracks seemed to be identical with the upper fillet crack of an actual rail described in Section 4.14.2.

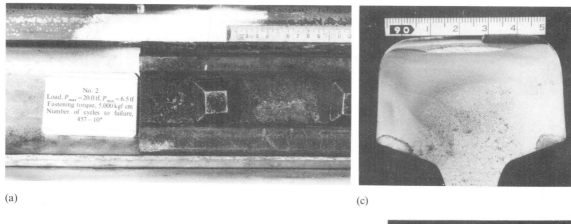

(a)

(c)

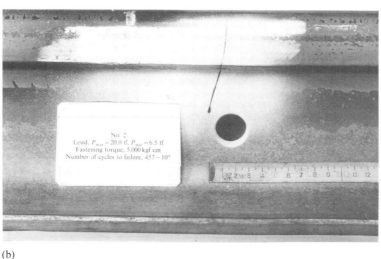

(b)

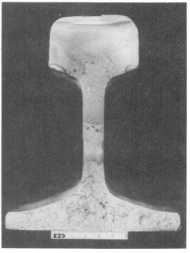

(d)

Figure 4.281   Reproduction test for upper fillet crack: (a) side view of upper fillet crack; (b) after removed fish plate; (c), (d) fracture surface of upper fillet crack

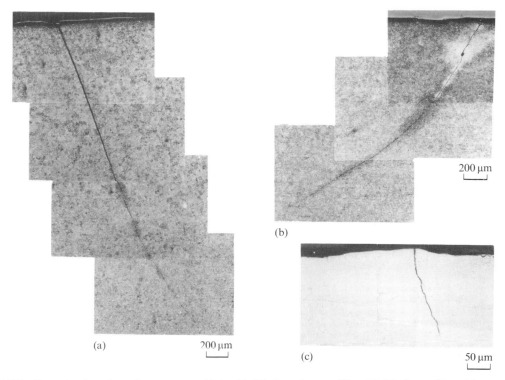

Figure 4.282   Example of cracks at the contact position with fishplate of upper fillet: (a), (b) after etching; (c) no etching

(b) Factors involved in the occurrence of an upper fillet crack

Fretting fatigue is heavily affected by the pressure working on the contacting surfaces, their relative slip, the coefficient of friction between the surfaces and environment atmosphere. Therefore, the occurrence of an upper fillet crack in the joined portion of rails is related to these factors and influenced by the force with which the bolts are fastened, the material of the rails, the weight of the wheels, and the condition of the ballast and rail lubrication. From the above influencing factors, those which can be readily controllable were chosen and studies carried out as described below.

(c) Influencing factors studied

1. Bolt fastening force
2. Lubrication
3. Type of rail steel
4. Residual stress

(d) Testing methods

Table 4.59 shows the testing methods employed.

(e) Tests results and consideration

(i) Bolt fastening force

The test results are shown in Table 4.60. Obviously, the greater the bolt fastening torque, the greater the number of repeated cycles to failure. This may be construed as follows: increasing the bolt fastening torque increases the surface pressure between the contacting rails and joint plate but decreases their relative slip. The results obtained as a whole occurred under the influence of the amount of relative slip.

(ii) Lubrication

Table 4.61 shows the influence of lubrication on the occurrence of cracks at the upper fillet of rails. Without lubrication, cracks occurred in the head-hardened rail. But application of lubrication resulted in an improvement in that the same rail withstood a repeated application of stresses at 1000×10^4 and over. This improvement may be attributed to the reduction of the coefficient of friction between the surfaces in contact with each other that was brought about by the application of lubrication, which, in turn, decreased the tangential force and prevented wear. Ordinary machine oil, even a considerably deteriorated one (containing 2% of water), proved to be effective. (No macro cracks were observed even after repeated loading of 1000×10^4.)

Table 4.59   *Experimental method*

Items for study	Rail	Fishplate	Fastening torque of bolt (kgf cm)	Test method
Fastening torque of bolt	50 N, HH	Normal type with 4 holes	Three levels: 2500, 5000, 7500	Figure 4.280
Lubrication	50 N, HH	Normal type with 4 holes	Standard, 5000	Figure 4.280*
Kind of steel	HH, NHH, as-rolled carbon steel	Normal type with 4 holes	Standard, 5000	Figure 4.280
Residual stress	HH, removed residual stress HH	Normal type with 4 holes	Standard, 5000	Figure 4.280

* Rust preventive oil (CRC2-26) was coated between fishplate and rail every 24 hours (every 72×10^4 cycles).
HH, head hardened rail, whose head is quenched and tempered.
NHH, new head hardened rail, whose head is slack quenched by air blower. Hence, the structure is fine pearlitic.

Table 4.60   *Effect of fastening torque of bolt on crack initiation at upper fillet*

Fastening torque (kgf cm)	Load (tf)		Number of cycles to crack initiation
	P_{min}	P_{max}	
2500	6.5	20	372.8×10^4
5000	6.5	20	450.0×10^4
7000	6.5	20	489.0×10^4

(iii) Type of rail steel

Table 4.62 shows the influence of the type of rail steel on the occurrence of cracks in the upper fillet of rails. Cracks occurred in both head-hardened and new head-hardened steels. But no crack occurred in the as-rolled carbon steel even after 1000×10^4 times repeated loading. This seemed to be due to the originally expected difference in the distribution of residual stress at the surface of the upper fillet of the rails. The smaller number of repeated loading cycles to failure with the new head-hardened steel in Table 4.61 seemed to be due to the shorter span that was chosen for the somewhat short test specimen. As a consequence, the load was increased to attain the same stress, which, in turn, increased the relative slip between the joint plate and the portion of the upper fillet of the rail. Figure 4.283 shows the residual stresses measured in different types of rail.

Table 4.61   *Effect of lubrication on the crack initiation at upper fillet*

Test rails	Lubrication at upper fillet	Load (tf)		Number of cycles to crack initiation
		P_{max}	P_{min}	
HH rail	No lubrication	6.5	20	450.0×10^4
	Lubrication with CRC2-26 every 72×10^4 cycles	6.5	20	No crack initiation ($\geq 1000 \times 10^4$)

Table 4.62   *Differences in crack initiation at upper fillet for different rails*

Test rails	Fastening torque of bolt (kgf cm)	Load (tf)		Total tonnage of repeated load ($\times 10^4$ tf)*	Longitudinal residual stress at upper fillet (kgf/mm²)	Number of cycles to crack initiation
		P_{max}	P_{min}			
HH rail	5000	6.5	20	6.075	+20	450.0×10^4
NHH rail	5000	7.3†	22.5†	5.320	+12.0	350.0×10^4
Plain carbon rail	5000	6.5	20	13.500	−4.0	No crack initiation ($\geq 1000 \times 10^4$)

* Total tonnage of repeated load $= (P_{max} - P_{min}) \times$ number of cycles.
† Full span, 795 mm.

Table 4.63 *Effect of residual stress on crack initiation at upper fillet*

Test rails	Longitudinal residual stress at upper fillet (kgf/mm²)	Load (tf)		Number of cycles to crack initiation
		P_{min}	P_{max}	
HH rail (As heat treated)	+20	6.5	20	450.0×10^4
HH rail (Stress relieving annealing)*	+2	6.5	20	No crack initiation ($\geq 1000 \times 10^4$)

* Heated at 530°C for 1 hour and cooled in the furnace.

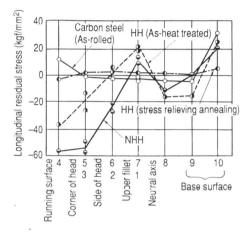

Figure 4.283 Longitudinal residual stress of rails (type of rail: 50N)

(iv) Residual stress

There was a residual tensile stress of 20 kgf/mm² in the longitudinal direction of the surface of the head-hardened rails (as heat-treated). When the rails were heated at 530°C for one hour and then cooled in the air (for stress-relief annealing), the residual tensile stress at the surface of the upper fillet decreased to 2 kgf/mm² (see Figure 4.283). Table 4.63 shows the results of a test in which the upper fillet crack of rails was reproduced. Cracks occurred in the head-hardened rail which exhibited a residual tensile stress of 20 kgf/mm² in the longitudinal direction at the surface of the upper fillet. In the head-hardened rail whose residual tensile stress was reduced to 2 kgf/mm² by stress-relief annealing, by contrast, no macro cracks were observed under the same test conditions even after 1000×10^4 times of repeated loading. As such, the residual stress in the longitudinal direction at the surface of the upper fillet of rails proved to have a great influence on the occurrence of cracks. Thus, reducing the residual tensile stress in the longitudinal direction at the surface of the upper fillet of rails seemed to be an effective way to prevent the occurrence of cracks in that region. (The stress-relief annealing produced little change in the hardness and structure of the rails observed under an optical microscope.)

The residual tensile stress at the upper fillet of the head-hardened and new head-hardened rails can be removed or changed into compressive stress by conventional shot peening or localized cold working with rollers. But these methods are not without problems in cost and ease of application. The most economical and effective way is to control the residual stress without significantly modifying the existing manufacturing process.

A method of controlling the residual stress in the new head-hardened rail will be briefly discussed in the following. For details, refer to cited pieces of literature [5,6]. It became evident that the condition of water cooling after slack quenching governs the set-up of residual stress in the new head-hardened rails. The large residual compressive stress at the surface of the head peculiar to the new head-hardened rails proved to be due to the conversion of a thermal stress set up by the water cooling applied after slack quenching into a residual stress. It was also found that the amount of the residual stress can be changed by controlling the rate of water cooling. Thus, it was discovered that the desired residual compressive stress at the upper fillet can be readily obtained by forcibly cooling (with water) not only the head of a rail but also the desired region below its jaw after slack quenching. Figure 4.284 shows an example of the results obtained. A test was conducted to reproduce the upper fillet crack of rails using the rail shown in Figure 4.284. As a conclusion, no crack was observed even after 1000×10^4 times of repeated loading (see Table 4.64). Further details will be reported on another occasion.

References

1. Kato, Y. (1978) *Rails*, The Society of Railroad Facilities, Tokyo, p. 271
2. Okazaki, A., Urashima, C., Sugino, K. *et al.* (1982) *Journal of the Iron and Steel Institute of Japan*, **68**, S1266, Tokyo
3. Urashima, C., Sugino, K., Nishida, S. and Masumoto, H. (1982) *Journal of the Iron and Steel Institute of Japan*, **68**, S1266, Tokyo
4. Waterhouse, R. B. (1984) (translated by J. Sato) *Fretting Damage and Preventive Measures*, Yokendo, Tokyo

Table 4.64   *Test results for crack initiation at upper fillet for controlled residual stress rail NHH (new head hardened)*

Test rails	Longitudinal residual stress at upper fillet (kgf/mm^2)	Load (tf)		Number of cycles to crack initiation
		P_{min}	P_{max}	
NHH rail	+12	7.3	22.5	350.0×10^4
Controlled residual stress rail NHH	−15	6.5	20	No crack initiation ($\geq 1000 \times 10^4$)

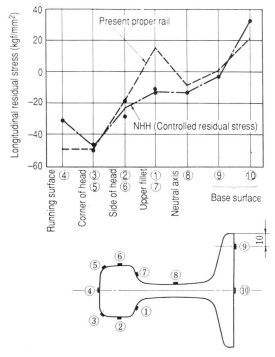

Figure 4.284   Longitudinal residual stress rails

5. Urashima, C., Nishida, S., Sugino, K. and Masumoto, H. (1982) Patent Application No. 223910-1982, Tokyo
6. Urashima, C. Nishida, S., Sugino, K. and Masumoto, H. (1982) Patent application No. 223911-1982, Tokyo

Conclusion

As mentioned in Section 2.4(c), several works dealing with the analysis of failures and collecting actual examples of failures have been published. In writing this book, therefore, care was taken to carry out studies from as many different viewpoints as possible from those of the predecessors. A feature of this book is that the discussion is detailed but easy to understand, with reference made even to universally applicable preventive measures.

This book uses the conventional system of engineering units (see the literature in Section 2.4(g)) because most of the instruments used at the production sites are still of the conventional type; there is therefore hardly any merit in using the International System of Units, and the figures expressed according to the conventional system of engineering units can be readily converted into those of the International System of Units by mental calculation with errors of only a few per cent. This conversion may be quite easy for college and university graduates and the young people who are going to work in the production equipment control divisions. However, people who have worked in those divisions for a long time may not be very familiar with the International System of Units.

Even so, the International System of Units is being adopted increasingly extensively throughout the world. This matter will therefore be reconsidered when a chance arises for reprinting. Until then, reference should be made to the literature cited in Section 2.4(g).

In the analysis of fractures, the applied loads are assumed to be simple loads with fixed stress amplitudes. Actually, however, many of the loads are probably of the variable type whose stress amplitudes change with time. Strongly recommended therefore is the literature on variable loads cited in the latter half of Section 2.4(g). A commercially available load waveform analyser will prove useful in making a more accurate analysis of failures. An analyser developed under the guidance of Professor T. Endo, Kyushu Institute of Technology, and marketed by Ono Sokki Co. Ltd under the trade name 'Stress Wave Analyzer' analyses the actual stresses to determine their maximum and minimum values, rain flow, stress amplitude and peak values.

Finally, the mention of three company names in this book is not meant for advertisement but for the convenience of readers.

Appendix 1 Conversion of units

Table A1.1 *Conversion table of units*

Unit	in, lb	mm, kgf	m, N
Length	1 in	25.40 mm	0.02540 m
	0.03937 in	1 mm	0.001 m
Force	1 lb	0.4536 kgf	4.448 N
	2.205 lb	1 kgf	9.807 N
Stress, σ	1 ksi	0.7031 kgf/mm^2	6.895 MN/m^2
	1.422 ksi	1 kgf/mm^2	9.807 MN/m^2
Stress intensity factor, K	0.9099 1 ksi$\sqrt{\text{in}}$	3.225 3.543 kgf/mm$^{3/2}$	1 1.099 MN/m$^{3/2}$
	0.2822 ksi$\sqrt{\text{in}}$	1 kgf/mm$^{3/2}$	0.3101 MN/m$^{3/2}$

Appendix 2   Models of fatigue crack propagation

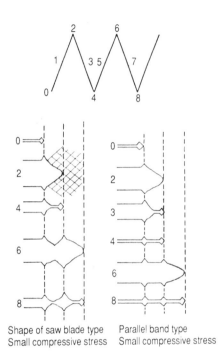

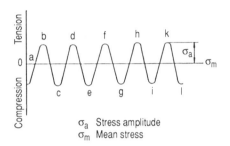

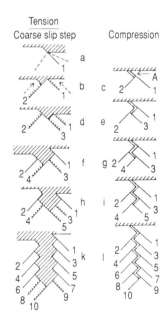

Shape of saw blade type
Small compressive stress

Parallel band type
Small compressive stress

Figure A2.1   Model of fatigue crack propagation [1]

References

1. Kitagawa, H. and Koterazawa, R. (1977) *Fractography*, **85**, Baifukan
2. Neumann, P. (1967) *Zeitschrift für Metallkunde*, **58**, 780
3. Neumann, P. (1969) *Acta Metallurgica*, **17**, 1219

Figure A2.2   Neumann's model of fatigue crack propagation [2,3]

Appendix 3   Notch factors for various kinds of rod specimen

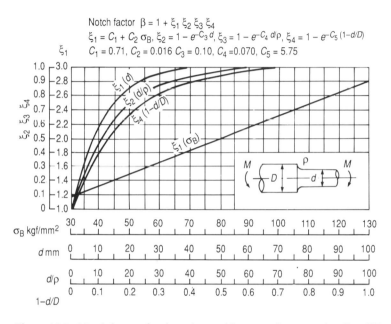

Notch factor $\beta = 1 + \xi_1 \xi_2 \xi_3 \xi_4$
$\xi_1 = C_1 + C_2 \sigma_B$, $\xi_2 = 1 - e^{-C_3 d}$, $\xi_3 = 1 - e^{-C_4 d/\rho}$, $\xi_4 = 1 - e^{-C_5 (1-d/D)}$
$C_1 = 0.71$, $C_2 = 0.016$ $C_3 = 0.10$, $C_4 = 0.070$, $C_5 = 5.75$

Figure A3.1 Notch factor of rod specimen with step under rotary bending [1]

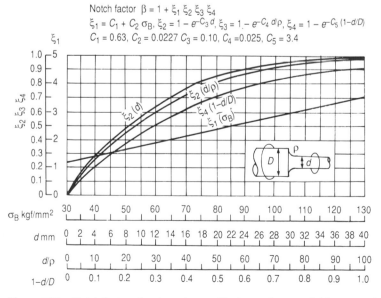

Notch factor $\beta = 1 + \xi_1 \xi_2 \xi_3 \xi_4$
$\xi_1 = C_1 + C_2 \sigma_B$, $\xi_2 = 1 - e^{-C_3 d}$, $\xi_3 = 1 - e^{-C_4 d/\rho}$, $\xi_4 = 1 - e^{-C_5 (1-d/D)}$
$C_1 = 0.63$, $C_2 = 0.0227$ $C_3 = 0.10$, $C_4 = 0.025$, $C_5 = 3.4$

Figure A3.2 Notch factor of rod specimen with step under completely reversed torsion [1]

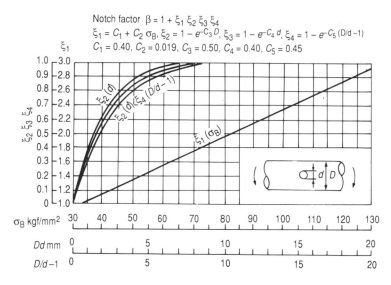

Figure A3.3   Notch factor of rod specimen with side hole under rotary bending [1]

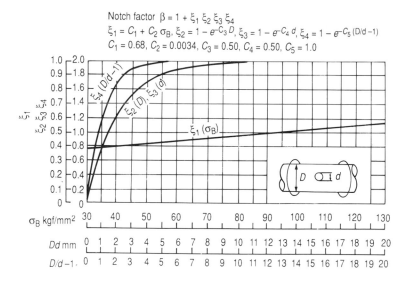

Figure A3.4   Notch factor of rod specimen with side hole under completely reversed torsion [1]

Notch factor $\beta = 1 + \xi_1 \xi_2 \xi_3 \xi_4$

$\xi_1 = C_1 + C_2 \sigma_B$, $\xi_2 = 1 - e^{-C_3 D}$, $\xi_3 = 1 - e^{-C_4 d}$, $\xi_4 = 1 - e^{-C_5 \{(D/d) - 1\}}$

$C_1 = 1.8$, $C_2 = 0.022$, $C_3 = 0.50$, $C_4 = 0.46$, $C_5 = 0.465$

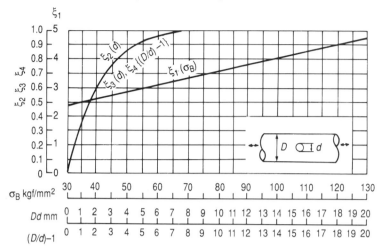

Figure A3.5 Notch factor of rod specimen with side hole under repeated tension and compression [1]

Notch factor $\beta = 1 + \xi_1 \xi_2 \xi_3 \xi_4 \xi_5$

$\xi_1 = C_1 + C_2 \sigma_B$, $\xi_2 = 1 - e^{-C_2 d}$, $\xi_3 = 1 - e^{-C_4 d/\rho}$, $\xi_4 = 1 - e^{-C_5 (1-d/D)}$, $\chi_5 = 1 - e^{-C_6 (\pi - \theta)}$

$C_1 = 0.57$, $C_2 = 0.0057$, $C_3 = 0.070$, $C_4 = 0.082$, $C_5 = 12$, $C_6 = 17$

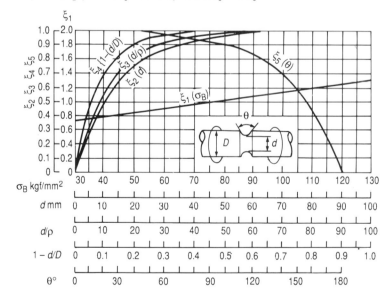

Figure A3.6 Notch factor of rod specimen with v-notch under rotary bending

Notch factor $\beta = 1 + \xi_1 \xi_2 \xi_3 \xi_4 \xi_5$

$\xi_1 = C_1 + C_2 \sigma_B$, $\xi_2 = 1 - e^{-C_3 d}$, $\xi_3 = 1 - e^{-C_4 d/\rho}$, $\xi_4 = 1 - e^{-C_5\{1 - (d/\rho)\}}$, $\xi_5 = 1 - e^{-C_6 (\pi - \theta)}$

$C_1 = 1.1$, $C_2 = 0.022$, $C_3 = 0.070$, $C_4 = 0.095$, $C_5 = 12$, $C_6 = 17$

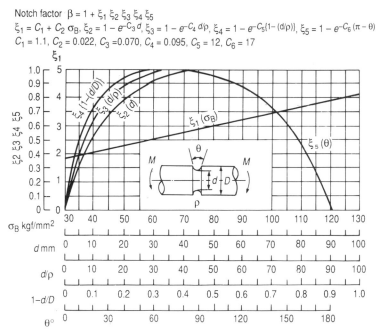

Figure A3.7   Notch factor of rod specimen with v-notch under completely reversed torsion [1]

Notch factor $\beta = 1 + \xi_1 \xi_2 \xi_3 \xi_4 \xi_5$

$\xi_1 = C_1 + C_2 \sigma_B$, $\xi_2 = 1 - e^{-C_3 d}$, $\xi_3 = 1 - e^{-C_4 d/\rho}$, $\xi_4 = 1 - e^{-C_5 (1-d/D)}$, $\xi_5 = 1 - e^{-C_6 (\pi - \theta)}$

$C_1 = 3.9$, $C_2 = 0.016$, $C_3 = 0.070$, $C_4 = 0.082$, $C_5 = 12$, $C_6 = 17$

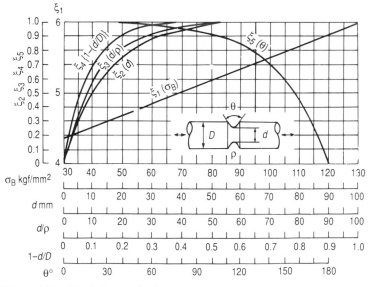

Figure A3.8   Notch factor of rod specimen with v-notch under repeated tension and compression [1]

Reference

1. Japan Society of Mechanical Engineers (JSME) (1982) *Design Handbook of Fatigue Strength of Metals*, Vol. 1, p. 125

Appendix 4   Typical fracture surfaces observed by SEM

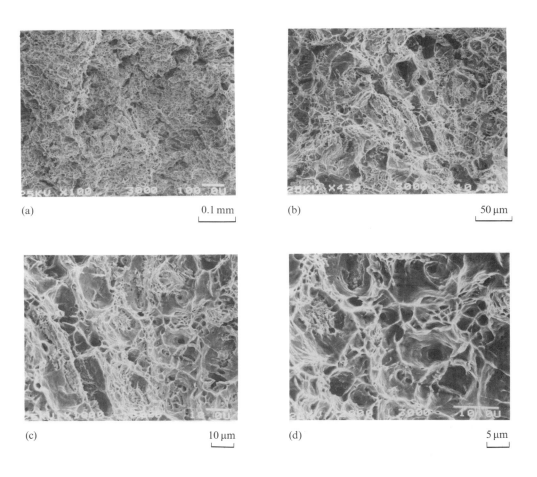

(a)   0.1 mm

(b)   50 μm

(c)   10 μm

(d)   5 μm

Figure A4.1   (a) Dimple fracture surface of Si–Mn steel ($\sigma_B = 50$ kgf/mm^2 grade). (b) Enlarged view of (a). (c) Enlarged view of (b). (d) Enlarged view of (c)

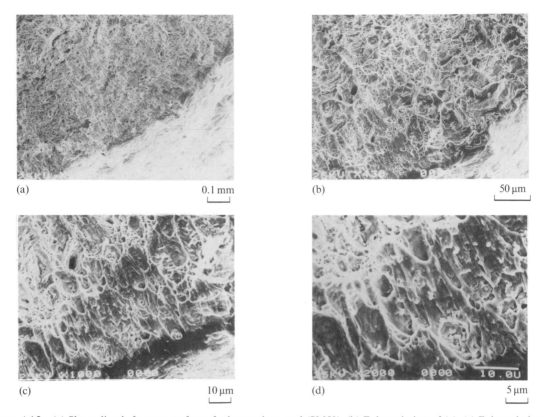

(a)   0.1 mm

(b)   50 µm

(c)   10 µm

(d)   5 µm

Figure A4.2   (a) Shear dimple fracture surface of a low carbon steel (SM50). (b) Enlarged view of (a). (c) Enlarged view of (b). (d) Enlarged view of (c)

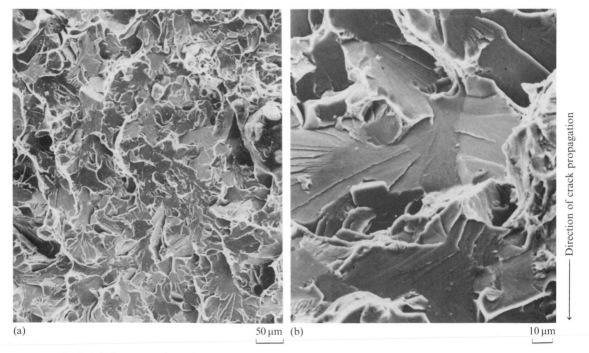

(a)   50 µm (b)   10 µm

Figure A4.3   Brittle fracture surface of a medium carbon steel: (a) typical cleavage fracture surface; (b) enlarged view of (a)

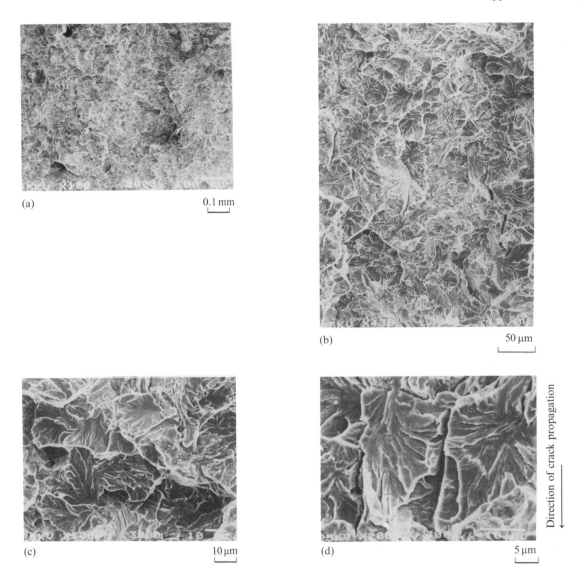

Figure A4.4 (a) Brittle fracture surface of a rail steel. (b) Enlarged view of (a). (c) Enlarged view of (b). (d) Enlarged view of (c)

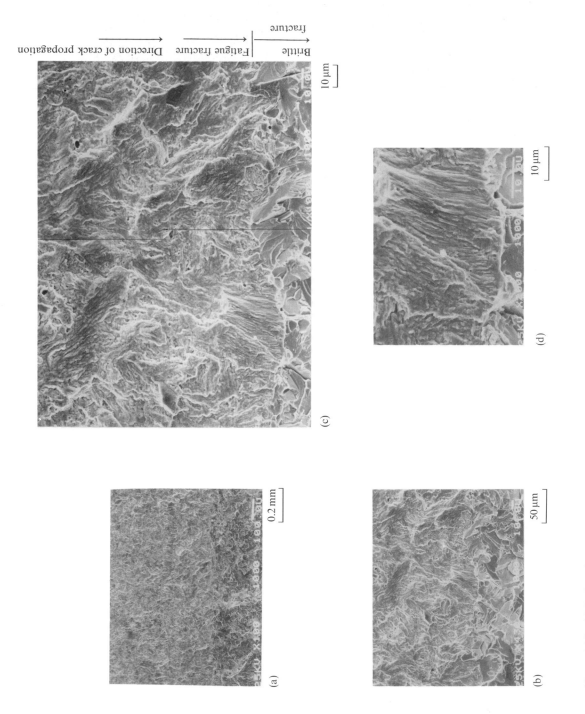

Figure A4.5 (a) Fatigue fracture surface of S45C (by rotary bending fatigue). (b) Enlarged view of (a). (c) Enlarged view of (b). (d) Enlarged view of a part of (b)

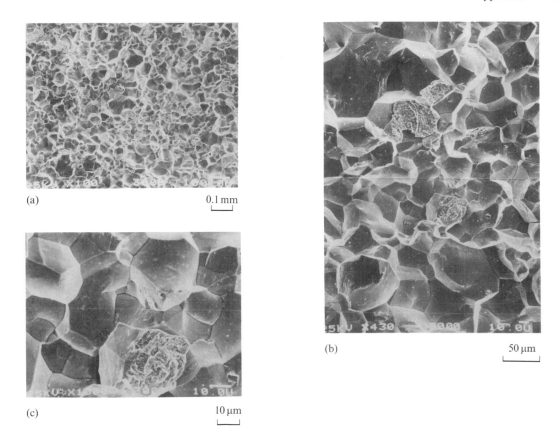

(a)   0.1 mm

(b)   50 μm

(c)   10 μm

Figure A4.6   (a) Quench cracked fracture surface of a rail steel. (b) Enlarged view of (a). (c) Enlarged view of (b)

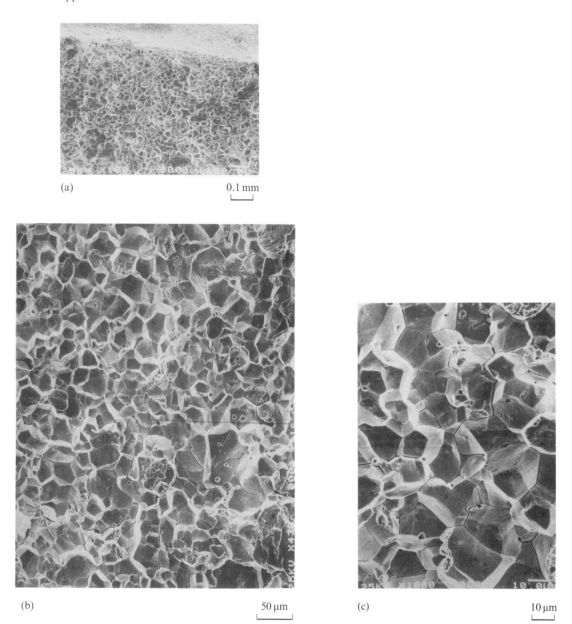

(a)   0.1 mm

(b)   50 μm   (c)   10 μm

Figure A4.7   (a) Delayed fracture surface of a high-tension bolt (F11T). (b) Enlarged view of (a). (c) Enlarged view of (b)

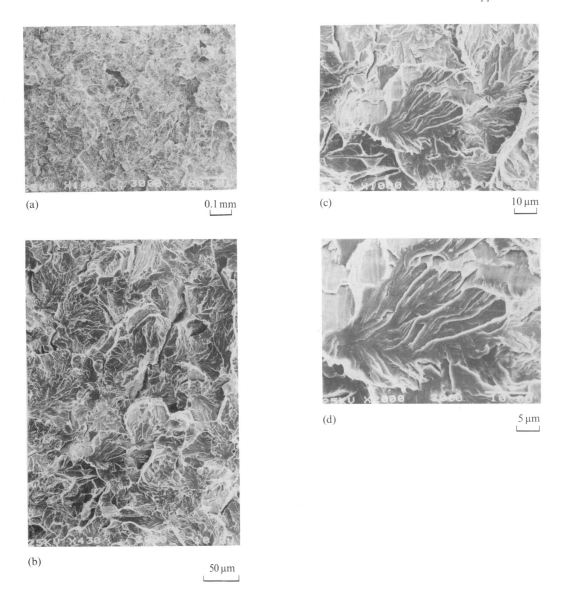

(a) 0.1 mm

(c) 10 µm

(b) 50 µm

(d) 5 µm

Figure A4.8   (a) Delayed fracture surface of a rail steel. (b) Enlarged view of (a). (c) Enlarged view of (b). (d) Enlarged view of (c)

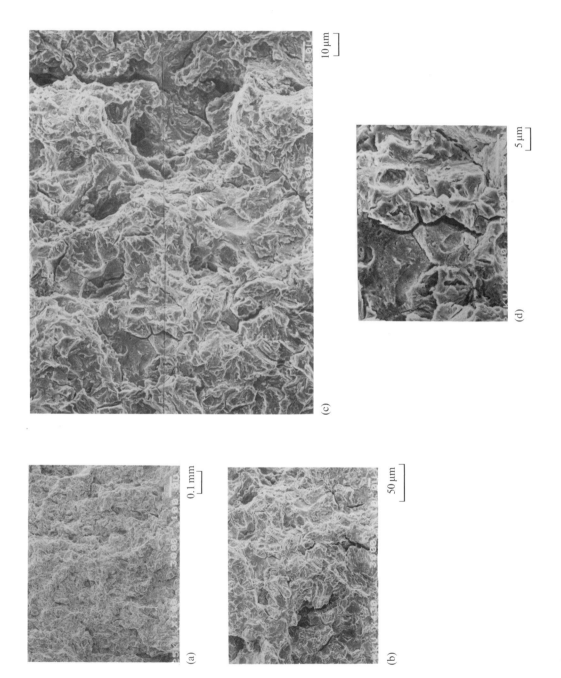

Figure A4.9 (a) SSC fracture surface of a C–Mn steel. (b) Enlarged view of (a). (c) Enlarged view of (c). (d) Enlarged view of (c). (d) Enlarged view of (c). (SSC: stress corrosion cracking, NACE TMO177, the solution comprising 0.5% acetic acid + 5% NaCl and distilled water with H_2S under 1 at a)

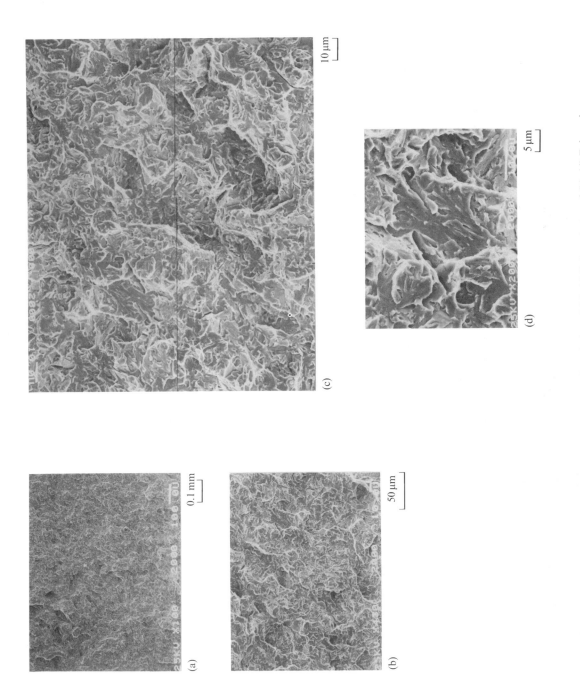

Figure A4.10 (a) SSC fracture surface of a Cr–Mo steel. (b) Enlarged view of (a). (c) Enlarged view of (b). (d) Enlarged view of (c)

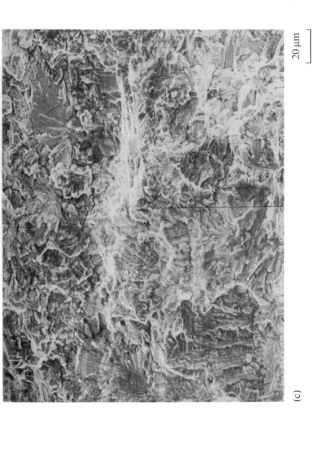

(a)

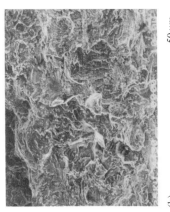

(b)

(c)

Figure A4.11   (a) SSC fracture surface of 22Cr–25Ni–4Mo steel. (b) Enlarged view of (a). (c) Enlarged view of (b)

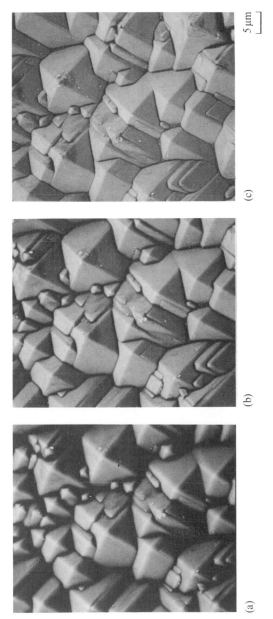

5 μm

(a)   (b)   (c)

Figure A4.12 Effect of accelerated voltage of SEM on the image of the surface of a solar battery (taken by auto-focus). Accelerated voltage: (a) 15 kV; (b) 10 kV; (c) 5 kV

Index